Arithmophobia

How to Heal the Horror of Mathematics

Cacildo Marques

Cover design: Cacildo Marques

Copyright © 2018 Cacildo Marques
All rights reserved.

ISBN: **978-1729719985**

Marques, Cacildo
Arithmophobia: How to Heal the Horror of Mathematics / Cacildo Marques. Maryland, 2018.

117p.
ISBN: **978-1729719985**

1. Mathematics Education. 2. Mathematics Anxiety. I. Title

DDC 510.71

CONTENTS

	Preface	vi
0	Prolegomena	1
1	General failure	5
2	Mathematics as a game	16
3	The passage	22
4	Basic operations	33
5	The fiasco of the Modern Mathematics	38
6	Correction in language	40
7	Peano's axioms	50
8	Strong morives	55
9	The strength of Geometry	83
10	The etiology	91
11	Redemption cases	98

Preface

In general the student does not confess that he is afraid of Mathematics. Anyone who has a problem with the subject says, "I do not like math."

Whenever someone told me this, I said, "Do you like to breathe?"

They found my question unreasonable. I explained: "Nobody wonders whether or not one likes to breathe, because breathing is vital." So I said that math is not as vital as breathing, but that is almost it. No self-respecting progress exists without Mathematics behind.

In my time in the classroom of basic education, especially in High School, I followed the downfall of the level of education and the consequent increase in rejection of Mathematics. I saw this refusal to learn the matter as a cultural problem, regardless of whether it could be interpreted as a psychic disorder. Researchers from the United States concluded that yes, it is a disorder, it is not only a matter of preference, like whether or not to like soccer, whether or not to like jazz.

It may seem that this information is bad news for those who do not get along with the Mathematics tests, but it is the other way around. As a disease, rejection of mathematics belongs to the list of those that are easily curable.

As a contribution to the theme, I present this book. In it I deal with the most basic operations of Arithmetic, without prejudice and without fear of being seen as a person who bends to leave, for a few moments, advanced Mathematics topics to worry about the fundamentals of the subject. Those who disregard this attitude simply yield to the crooked view of vulgar materialism. In the medical field, the choice between being a pediatrician or being a geriatrician does not diminish or elevate the professional. In didactic, many think that what works with children and adolescents is diminished, but such an interpretation is based only on the difference in wages. If Ludwig Wittgenstein did not care for small minds, when he dedicated his life to teaching children, neither should we be shaken by the arrogant ones.

Condorcet, as the reader will see below, gave me the inspiration to develop the approach I present here. If we are prejudiced against elementary Arithmetic, then we are fueling the horror of Mathematics by the victims of the disorder. To overcome this and so many other problems, it is necessary to open not only the eyes but also the mind, at the same time.

Cacildo Marques, November 2018

Arithmophobia

How to Heal the Horror of Mathematics

Cacildo Marques

Arithmophobia

How to Heal the Horror of Mathematics

Arithmophobia: How to Heal the Horror of Mathematics

0. Prolegomena, ha-ha!

Let us address the "etiology" of the problem a little further, but here is the importance of looking at the fact that, as everyone knows, there is a focus of arithmophobia based on the Iberian culture, which reaches Portuguese-Hispanic and also descendants that grow in cultures as different as the German, the Hungarian and the American.

As I did in the book "Flexible High School", of 2017, I use the first person here, unlike any other essays I have published, because we are on the ground that involves a lot of personal experience, and the detachment of the third person would bring some hang-ups in the interaction between me and you, reader, on the subject that needs a conversation as informal as possible.

In a meeting on Mathematics Education that I promoted with colleagues when we were graduation students, I presented in a speech the question of the difficulty of teaching Mathematics in a country that inherited rejection by the subject, which produces in adolescents the psychic problem that in the United States was named "mathematics anxiety", a precise expression that, after all, does not account for the whole dimension of the disorder. We should not be prone to classify the thing as psychic disorder, once we know that some types of disorders are culturally inherited.

The experience of those who have spent years teaching Mathematics shows that such "anxiety" comes from the refusal to absorb arithmetical content, so that solving the problem of rejection of numbers in the young student cures the "horror of Mathematics" in general.

The advantage of suffering from arithmophobia in the face of the misfortune of carrying more serious psychic disorders is that this "sickness" of which we speak here is easily dissolvable without the need for medication, psychoanalysis or punishment. The reference to this latter method is not surprising, since, until 1907, when Maria Montessori founded in Rome the Casa dei Bambini (House of Children) school, it was used in school and home without questioning, except by one or another rebel student.

Description. Let us make a brief diagnosis of the case.

Before, it is important to take into account that arithmophobia grows among children in the direct reason of the abandonment of the teaching of Geometry in the school system, mainly in the public system. With this I am anticipating, strategically or foolishly, a clue about the method of treatment.

If there is a good teaching of Geometry - if possible, also Music - we have an indication that education is being cared for with great zeal in the country, or in the province, in question. And if it occurs competently from the earliest school years, it will be rare the case of the student "carrying" the disorder of arithmophobia.

When the Geometry learning base is rare or absent on the part of the students who have to study, for example, the High School Algebra, there is a substantial number of adolescents affected by math anxiety.

On the eve of the Mathematics test the student remains in a state of tranquility, even though he knows that, without having had the courage to study, he will have a very bad result, if, in fact, he tries to solve the questions that the teacher presents. The parents do not know that the next day there will be this test, otherwise, applying responsibility in the adolescent, they will simply bring up the manifestation of the problem beforehand, although in a different way from that which will occur in front of the master. One possibility is that the student is affected by dysentery, either the day before or a few hours earlier on the day of the test.

If there was no previous pressure, the student enters the classroom without any psychological shock. The dread arises when the teacher enters with the package of tests. Or when he start writing the questions on the blackboard, if this is the case. It is something like what feels the very shy individual who has to take the stage and make a speech. There, the person makes the speech, with the risk of stuttering, choking or losing his voice, but physically nobody notices anything weird. With arithmophobia, no. As soon as the student sees the exam, or sees the first question started on the blackboard, or even a few minutes before, the color of the skin changes. It is like the blood is coming out of his veins. He can either turn white almost like a sulphite paper as he can get a greenish color. As there is indeed a change in blood pressure, he realizes that he is feeling sick. If the teacher insists that he take the test, he can complicate the situation. The most sensible thing to do is to consider that the student is not able to continue his academic activity at that moment and postpone the application of the test. He must, of course, talk to the student, reassuring him, trying to get his fear out of all that developed content. The fact that he realizes that he will have more time to prepare, if he even addresses the situation, already serves as a foothold.

The ideal thing, of course, is for the student to be healed, and do not new cases arise.

If you, reader, are seeing some exaggeration in the above description, since you must know people with arithmophobia who are free from physical disarray, I assure you it is not exaggeration. This is the most advanced form of the disorder. The young one who is going through badly before the vision of the test that he should do, can go through the need for hospital care. It should not be understood as a simple situation, always surmountable with a little sugar water to calm.

The situation of the student who shows no change in the physical aspect, nor of change of color nor of tremor, nor of any other aspect, but who cultivates a serious rejection to Mathematics, like those that systematically deliver the test in white, saying "I know nothing" or "I do not like this matter", this situation is much more widespread. These young people have self-control, built on the perception that they are part of a chain, of those who "hate Mathematics". It is as if they were part of the crowd of a sports team, who psychologically strengthens and supports each other member. And there is, of course, the student who does not even know of others victimized by arithmophobia (very rare case), but who has his self-control, and does not do "drama" before the test.

The understanding that the arithmophobe makes "drama" is equivalent to that of the family member who imagines that his relative with depression problem is making gender. It is obvious that there is drama, but it is not a matter of pretense or exaggeration. It is a problem, both in depression and arithmophobia.

There is a complicating in the follow-up of student cases prone to developing the arithmophobia received from the social environment, and this is the fact that the difficulty he has with learning can transcend rejection to numbers. Here, the dark cloud is diluted by observing student performance in other reasoning subjects, as a foreign language, provided that the assessment in this matter follows the normal curve, rewarding scholars and "punishing" the "defaulters". If the teacher of the comparison matter does not take the evaluation seriously, assigning high marks to the scholars and also those who sign other people's work, then he discards this one and searches for another. It can be the mother tongue itself, if the evaluation in it is efficient, attentive. Otherwise, the area of natural sciences is used - in the case of High School, Biology, for example, that does not depend very much on numbers.

Arithmophobia: How to Heal the Horror of Mathematics

1. General failure

If the student has poor overall performance and can only obtain a grade via fraud or in those subjects whose teachers are patronizing in the evaluation, then we are faced with a student who cannot learn a foreign language or the interpretation of a musical score. He is a student who reads a paragraph of fifteen words and cannot know the subject matter there, although the latter fact should not be used as a criterion, as we will see later. His father should not expect that in Mathematics he has great success, although the possibility exists.

The student may be experiencing a phase of absence of "animal spirit", of complete lack of enthusiasm. Performance will be low in the general subjects because he does not engage. It is in a state similar to the victims of "chronic fatigue syndrome". It is as if he suffered from logophobia, horror to reason and the word in general, as well as not interacting with gymnastics and sports. In this type of situation, seeking to cure arithmophobia may be a path, since Mathematics "works miracles", but one should not put much hope there. A more comprehensive treatment should be sought. (If the student began to use narcotics when a child and his brain got damaged, then there is nothing to do.)

There is a possibility that a low performance student will change his life course by changing attitude toward Mathematics because this subject matter is a behavior moderator. A student who is accustomed to studying Mathematics acquires predisposition to engage in other subjects. If he indulges in Mathematics and the exact sciences while harboring contempt for his mother tongue and the foreign language of his curriculum, this is almost the fault of the school system.

A student who performs well in languages and does poorly in Mathematics needs to have his case analyzed, being able to be a victim of arithmophobia or someone who has been taken by others to just devalue matter (i. e., proto-arithmophobia). In any case, he should receive guidance or treatment. Likewise, the optimum student in Mathematics with systematic low performance in languages should be guided and encouraged to devote himself to reading and Grammar. The student can learn to solve enormous equations, but he will fall in the face of discursive problems, which he has to decipher and model. Modeling, it should be remembered here, is the art of transforming a discursive utterance into an algebraic statement, that is, extracting the formal content from the text, thus mounting the algebraic and algorithmic structure of the problem to be solved.

Thus, for the student to do well in Mathematics, but also in Physics, it is necessary that before he has dedicated himself to Grammar and reading. Otherwise, his "mathematical" reputation will last little.

The student does not, however, need to be a voracious reader, of those who exchange books in the library each week. The condition of reading without difficulty and having a minimum of training in it can guarantee the basis for a good performance in Mathematics and Physics. He just needs to get used to the verbiage of these matters. If the learner has read so little that he has difficulty interpreting words and pronouncing them fluently, stammering in the reading without suffering from stuttering, and interrupting it by obvious lack of habit, he will have difficulty in any subject that requires writing, even if here or there he obtains high marks, in inconsistent evaluations.

Evaluations. Already started above the discussion on levels of evaluation as to efficiency or not, it is a good idea to look at this subject a little.

There are so many means and styles of application of tests and works, and final consolidation of the result of evaluations every two months - or every quarter, if this is the case -, that the family should not simply take the mark, value of the grade, as accurate diagnosis.

In tyrannical regimes, of the old dictatorships, the citizen, or subject, was, in many cases, submitted to the same ruler during his entire life. At school, the situation corresponded to the configuration of the policy, and the student had to "endure" the same master, for years, until completing his apprenticeship in elementary school. In republican democratic times, rulers are replaced periodically, and, in school, the student also passes through several teachers, both for the succession of the years and for studying under a rule in which different teachers develop different subjects within the same school year, at least from junior high school, if not already, in some cases, in the initial two triennials. This allows the student to compare different forms of evaluations among the different masters.

An insightful student will notice in a short time that there are subjects that he knows little and in them he has high grades, while there are others that he knows a lot and in these he has low grades. These discrepancies are intrinsic to the model of the modern, multitudinous school. And, contrary to what can be imagined, it is instructive for the student the perception that the teachers cannot maintain coherence among themselves in the assignment of marks.

There are academics who attack the numerical grading system based on this discrepancy argument. They do not realize the advantage of the model. If, by some magic of fate, all teachers were able to evaluate in an absolutely equal way, in a way that the student who learned 60% of the subject of the bimester in Grammar and 70% of the Mathematics subject got a grade 6 in the first matter and grade 7 in the second, and there was never any inconsistency in these measures, with the feeling of learning on the part of the student always corresponding to the grade received, then this situation would bring to the learners the implicit information that the authorities are unquestionable, omniscient and, as a consequence, tyrannical, now by total mastery of their "metier", unlike the ancient tyrant, who was naked for good eyes, and had to deceive the myopes, who were the vast majority of the commanded.

The inconsistent, incoherent grades between one teacher and another, serve to show students that their teachers, like their parents and their rulers, are human, without many agreements between them. They are people who often hit, but miss, just as children do.

With this little possibility of matching the grades assigned by the different teachers, the student is left to realize that the validity of the numerical grades is in the internal coherence of the work of each teacher. Suppose a teacher who in the first trimester values students who write in their own words, based on their own deductions, and who in the second trimester he gives good grades to students who have only memorized contents and reproduced them in the test. In the third trimester, this same teacher decides that the well evaluated will be those who can summarize their answers to the maximum, compressing them in one or two lines, and that in the fourth trimester only gives high marks to those who were very analytical, and answered each question using a minimum of twenty paper lines. All these variations of evaluative method this teacher adopted without giving any clue to the students. What internal coherence does he have in his work? Students will realize that he does not have any.

Because this is absolutely unusual, teens know that every new teacher has his style and his requirements, and the student's job is to identify this and act on the expectations of his evaluator. It is clear that the frequent exchange of teachers with different styles disrupts students' learning and performance. If a new Mathematics teacher comes every trimester, each with his method, the students will suffer the bumps.

We have cases where student X, of the eighth year A, having teacher Peter as his Mathematics teacher, arrived at the end of the school year with average 9, not knowing much, since Peter is not very demanding in the evaluations, at the same time in which the student Y, of the eighth grade B, whose teacher of Mathematics, very demanding, was teacher John, arrived at the end of the school year with grade 6. As X and Y used to study together throughout the school year, because they are neighbors, both knew that the domain of content by Y was much larger than that of X, and Y was not the type to draw "blank" in the test. By the way, looking at the evidence, both also found that X made many more mistakes than Y. Now, how is this mismatch between evaluators possible? It is possible because teachers Peter and John have autonomy and have different ways of working. The law guarantees this. It is the freedom of professorship. (A problem of this freedom, which is a big problem, is that situation where Professor G develops one chapter of content throughout the year while Professor H, from another classroom, develops eight chapters. Of course, students of G are in frank prejudice of learning compared to those of professor H, unless this one runs with the subject without commitment with the accompaniment by the adolescents, which is something absolutely unusual, because to fulfill the content already is a form of demonstrate compliance with the expectation of quality teaching.)

Much more to avoid difference in content development than grade scale difference, some education systems establish unified tests, either in full or in part. By total unification, teachers develop content in the classroom, each in their own way, but the evaluations are the same for all classrooms. For example, if there are six eighth-grade classrooms with three Mathematics teachers, each teaching classes for two of these six, there will come two unified bimonthly tests if the rule is two tests over the two-month period. Both eighth-grade A and eighth-grade B students have to answer the same questions, both in the first month and in the second month, because they are unified tests. More than that, and for that determination to work, regardless of the slowness or haste of each teacher, some systems use separate two-month handouts. Completing the first two months, the handout is collected and the students receive the new handbook, for the second two months. If a teacher has failed to complete the subject of the previous handout, which has been collected, he cannot continue working on those final topics in the new two-month period. Students have the new handout and they will have to work on its content. The very slow teacher, who had the habit of working "exhaustively" the contents in the beginning of the bimester without ever completing the chapters, should correct his attitude or deliver the job vacation to another.

This scheme of unified tests and handouts collected at the end of the two-month period works well, but it hurts the freedom of teaching in some way. Only private schools or schools that, within the public network, receive authorization to develop a differentiated project, can maintain the model.

The ideal is partial unification. Teachers can apply tests and assignments throughout the two-month period, but agree that one of the tests is unified. In this case, if it is an official school, teachers can prepare the test together. In private school, as usual, this unified proof can come "from top to bottom". With a unified test per bimester, the school system is able to gradually lead teachers to work in common rhythm, without that situation of teacher G in comparison with teacher H of the example given above.

If the establishment's teachers are committed to improving unit performance, and if the school management embraces the philosophy of quality, there is no reason to resist the establishment of partial unification of assessments. This can be done in junior high school, High School and college.

Numerical. In any case, students should be instructed early on the fact that the numerical grades they get in evaluations are relative measures. The grades assigned by a lecturer should preferably be close to the grades given by other lecturers of the same subject, but they will not necessarily maintain terms of comparison. More distant still will be marks of distinct schools, of districts with social levels also many different each other. The university systems that take into account in their selective processes the grades of the students' schools commit a serious error of interpretation of facts. Marks from different schools are only very grossly comparable. A mark 7 (or grace C+) of an MIT Calculus test is unrelated, except for the number form, with mark 7 of a Calculus test of the same chapter of the engineering college of the city of Foot of the Mountain, which charges a monthly fee of five dollars.

Is this difference in level between school units a big problem? Not at all. If some authority says that all units in his school network have the same level, this is a sickly liar. If he says that they will have this equality at the end of his management, by goal, then besides being a liar, he is a land seller on the Moon, a misleading.

It is because they have differentiated levels that school units can compete with each other. And healthy competition is the key to progress. Whoever says that it does not compete for improvement, because it does not accept to compete, is competing for the worsening. Competition is in the reptilian brain of animal species and those who do not adopt healthy competition regress to the gross competition of natural selection, and this is what happens when the competition itself is denied. The gross competition of natural selection is absolutely unconscious. Healthy competition, unlike it, requires clear rules, mobility, and discipline.

Once the numerical evaluations of distinct units are not comparable, does this mean that a student with a history full of red marks below 5 may be more studious and skilled than another student with a history full of blue marks? It can happen, yes, but let us be calm. If a student has a history full of red grades, these results refer to the evaluation system of his school unit. Within it, he is a low-performing student. If he moves on to a less demanding unit, he can change his life. The little that he studied at the previous school, and that did not pay him a grade, in the new unit may be enough for him to draw blue marks.

With such inconsistency in evaluations, would it not be better to abandon numerical grades and adopt mentions in letters, or just to adopt opinions?

The answer is no.

As much as numerical grades are suspect, mentions or opinions are incomparably more likely to represent acts of will. Without counting, the opinion of a teacher about a particular student may be much more related to the color of the eyes than to the performance accumulated over the two-month period. If the teacher tries to be minimally fair in the evaluation of his students, he must compute their performance, and turn the values into a final grade, preferably from zero to ten.

In a grades system from zero to ten, students will make their own performance accounts over the course of the bimonthly period and throughout the school year. In a system of mentions, they have to wait for the teacher's verdict at the end of the process. With accounts in the numerical system, students will have reason to learn to do the arithmetical operations. By themselves, they learn to calculate averages, because they have an interest in it.

Arithmophobia: How to Heal the Horror of Mathematics

The United States, which have devoted a great deal of effort lately in the struggle to improve their basic education system, will gain a great deal of encouragement if they adopt numerical grades in all states rather than mentions in letters.

Are there justifications for a system to adopt mentions rather than numeric marks? There are, but the greatest of them is unconfessable, because it comes from the arithmophobia. In postgraduate systems, in colleges, the custom is to assign mentions in letters. The reason is very clear: free will. Unlike undergraduate, which uses numerical grades, in masters and doctoral courses it is well-known that professors and counselors retain power of choice because they will have to choose who will be part of their working community. Between two students, one can obtain grade 10 and another, grade 8, but, by subjective evaluation, professors may conclude that the second is preferable to the first one, because it is more creative and more adapted to academic life. In this case they can give a mention B for the one of mark 10 and a mention A, superior, for the one of mark 8. Of course the situation seems unfair, but it is not strange. It should be remembered that the universe in which this occurs is very restricted, of hundreds of people within a country. Giving this arbitrary power to teachers who work in elementary school, caring for many millions of students each school year, is not at all wise.

All this discussion was generated by the realization that the fact of taking a red mark in Mathematics at school does not necessarily mean that the student is a fatal candidate in the gallery of the victims of arithmophobia. It may happen that he has relaxed in his studies, because he is dating, or because he has spent much time on the Internet, and there may have been an incoherent evaluation by the teacher as well. He may also be attending a school too demanding for his behavior pattern. In this case, parents need to be careful when looking for a more condescending school, because some schools are so uncompromising with evaluations that a student who learns nothing new throughout the school year is getting blue grades, fooling himself and his parents.

How to know if the student is doing too badly by just looking at his red grades? By comparison with the rest of the class, because his teacher's grades have internal coherence. So, if Johnny took mark 4 in the second trimester, but half of the class scored below 4, Johnny is fine. Otherwise, if he took 5, a blue mark, but all classmates took more than 5, so the case of Johnny is worrisome. In a more demanding school, or in the hand of a more demanding teacher of the same unit, he would have a grade 2, for example.

Memorization. One possibility that is almost non-existent today, but not totally disposable, is that of the teacher who demands everything by heart, memorization by memorization. In this case, the boy can be a brilliant student, who deduces the whole multiplication table and solves complicated problems, and nevertheless gets very low marks with the unfortunate teacher. What happens is that the student does not accept to spend time memorizing, once he has great skills of deduction, and will have low marks, because the test is not really evaluating him, for wanting from him something that in the 21st century does not make sense.

But the family should also not buy the destructive speech that nothing should be learned by heart. This conversation is damaging. I do not have to memorize as much as 4 times 7 if I know the result of 2 times 7, 14, and I can deduce the other result quickly, mentally calculating double of 14 (since multiplicand 4 is twice the multiplicand 2). And the table for 2 I learned it by knowing that each line number is equal to its sum with itself, being, for example, 2*3 equal to 3+3. Thus, much in school does not even have to be learned by heart by those who know how to deduce. But the resolutive formula of the equation of the second degree, denominated by Professor Castrucci as Formula of Baskhara, is not something of fast deduction. If the student is going to take a high school equation test, he knows he should not let the formula be deduced on time, because he will lose a lot of time and risks not reaching the final formula for some mistake in the middle of the way. It is clear to him that the best way is to have the formula by heart - the misguided student will look for some fraud, but this one does not go to school to educate himself.

Is the proportion of students who can deduce the multiplication table large? No, it is not. For a band of at least 90% of the pupils, the way is to learn by heart. Until the beginning of the twentieth century this was done by means of ruler blows, punctures of slapper, punishment and reprimand, because there was a short term and the rule, before 1907, as said above, was memorization by memorization. Whoever did not memorize was considered a bad student.

Are those 90% weaker students, since they have difficulty with deduction? It is not about this. We can classify these types of students as *heuristics*, those who deduce with ease, and those who are *sensitives*, who learn more by repetition and memorization. A heuristic student and a sensitive learner may have the same performance in a test under the same teacher, and then no different merit will be sought from those who have deduced or memorized. The classification between heuristics and sensitives is close to that of Joy Paul Guilford (1897-1987), who divided the students into those of *divergent* reasoning (as to the tests) and those of *convergent* reasoning.

A sensitive student memorizes the entire table, preferably through use, bit by bit, and then assimilates the algorithms for the operations, just as the heuristic student must also do. In a multiplication test, the sensitive student, who already has the table by heart, can finish the task before the heuristic student, who will have to spend some time deducting numbers.

I appeal here to my fellow teacher. Since at the beginning of the twentieth century the great mass of students had to leave school in the first three or four years, they had to memorize the table soon, for the sake of time. In the 21st century, the vast majority will have to remain in school for 11 or 12 sequential years. Those who cannot deduce the tables will have to memorize it. But now they have plenty of time. We can establish that by the middle of the fifth year (ten-year-old students), all students who have not yet had the multiplication table in their head can consult it in the tests. It is not an ugly thing for a ten-year-old to have decorated the whole table. But it is an old, perhaps anachronistic attitude, to require him to memorize the table at an early age so that he can show skill in the accounts. In these times of electronic memory, much more important than knowing the table by heart is knowing how to do the operation in pencil.

Understanding. Let us take the example of multiplying the number 7859 by the number 48. If the student set up the account and made the operation, with the memorized table he was expert, but consulting the table beside him he also showed that he learned, that he knows the multiplication algorithm. Obviously, one of the resources of the teacher is to offer punctuation to those who do not need to consult the table in the exams. For example, if you consult, the test is worth 9, worth 10 for those who do not consult. It will be a means of pressuring the class to get rid of the table. But the ideal is that it should be abandoned for losing grace, just as the student abandoned the pacifier years before. Obviously, it should lose its grace when the student knows all about it.

The parents who accompany their children in the studies should also take into account that the most gentle way to memorize something is to do this by use. The table has the advantage of always repeating itself, if the student is doing the operations. If a child as young as six gets into the habit of doing math exercises every day, he will almost certainly have the multiplication table memorized before the age of nine.

Later we will study a Japanese table, made on the fingers, like a type of abacus.

The student should be instructed at an early age not to think that it is enough to "understand" a content so that this content is already considered learned. The term "to understand", in fact, is used inappropriately by children, because what they understand by "understanding" is "knowing what it is". When we present something to someone, we introduce a person, for example, that someone "knows" the person. Living with him, we assimilate him. In time, we can come to understand him.

In the case of Mathematics, the first step is this "knowing what it is", the first contact. The student is "knowing" the content. In the exercise phase - in fact, many exercises -, he is "assimilating" the subject. When the assimilation is very advanced, then he can have confidence that he is "understanding".

He should not expect to understand then to start doing. He should know, yes, the mechanism, even though he has no idea what is behind it. When a driver learns to drive a car, he learns the steering mechanism, not knowing how the engine, the main part, works. Anyway, he learns steps. Likewise, the student learns steps from mathematical operations, to begin doing them.

To do an addition operation, for example, the learner takes advantage of the sum table without very clearly understanding the mechanism of the addition algorithm. It puts unity under unity, ten under ten, and so on. It begins to add, from top to bottom. If he then inverts the plots and adds - proof of addition - using the same algorithm, he will get the same result. Intuitively he knows why this is working. There is no need to fully understand the algorithm at this point in his life.

Fingers. Nowadays the kids do not consult the sum table, but they operate on the fingers. To make 8 plus 5, the child starts from 8 and adds, one by one, 5 more fingers, getting 13. There is no harm in that. He is mastering the process of addition and is doing exercise with his fingers. While doing exercise every day, in a few months he will no longer need to use his fingers to make accounts.

In the phase of subtraction, the child may return to using fingers, but must learn to count backwards. To do 7 minus 4, for example, he sets the number 7 and counts on the fingers, after exposing 4 stretched fingers, 6, 5, 4, 3. Ready, he got the result!

To take the proof, he will add the subtrahend 4 to the result, which is the rest 3, to get the minuend, which is 7. In that he returns to practice the sum. In learning, operations are therefore recursive. If he has not learned the sum very well, when doing subtractions, the sum will come back. When he is doing multiplications, then the sum will also return. And in the division he will have to practice subtraction again, and also the multiplication and the sum when he takes the real proof.

When learning to read, one must learn letter by letter, so that reading appears. I met a boy who was at school three grade already and could not read - a colleague of him told me so. I went to check and saw that he knew all the letters, except one, the "R". By a letter, the whole mechanism of reading was impaired. Making him memorize the meaning of "R", I saw that soon he began to read normally. In the case of the number tables, it is not necessary to exhaust all the learning to move forward, since the student will always return to the previous topics, recursively. Obviously, he will have to learn the ten digits to begin with. From then on, just do not get discouraged.

A common feeling to all who receive a math test to do in class is that the entire subject is unknown. This is also true with test of foreign language, if it is well elaborated. But the math test can bring that feeling more often. When students have studied little, there is a temptation to deliver the test blank, early on. The experienced teacher receives no blank test before a period of at least half an hour has elapsed. If you, reader, have already experienced this feeling of not knowing anything when taking a test in your hand, forget the idea that you were less capable. If the test is poorly worked out, with questions requiring only memorization, then the student immediately recognizes what is being asked for. This type of exam comes with the name of Mathematics, but it is not Mathematics that praises itself. With a test of new questions, within the studied subject, the student feels lost at the first glance. Then he begins to situate point by point. If he has studied, he will now begin to knowingly resolve the issues. That feeling of discomfort disappears in the first few seconds.

2. Mathematics as a game

We are not dealing here with playing the Hash Game, or the Hopscotch. Mathematics exercises are games per se.

In more advanced problems, there is usually a response that the book presents in the final pages, or that the teacher already has beforehand, to later check the work of the students. In the simpler accounts of the four operations, these answers also exist, but the learner can check the outcome of the game without looking at the answer, but simply giving the real proof.

A neighbor once called me to see the case of her son, who had his diary full of division operations to do and did not even begin. He was taking low grades at school.

- Do you have time to see the case of my son Little Paul, who is doing poorly in Mathematics?
- Sure, I do, yes. What grade is he in?
- Fourth grade, but he risks repeating.
- I'll be there this Saturday at 3:00 pm. Can be it?
- Yes, you'll be welcome.

When I saw that quantity of divisions in his notebook, with no beginning, the first thing I asked was if he had learned to take the real proof. He had not learned. He said that the teacher had not yet taught this - he may not have paid attention, or was really speaking the truth, we will never know.

I took one of the accounts and showed him how to do it, also showing how to take the real proof at the end. Then he made the second, then the third, always taking the proof. I saw he got the mechanism. The following day I asked his mother if he continued those operations. She said he had done them all, which were over fifty.

What lacked for him was knowing how to take the real proof. He had the games in front of him, but he could not check whether he was "winning" or not. Knowing to take the real proof, he saw that he could win them all.

The moral of the story is that when we teach someone to go, we should always teach him to come back. If we teach just to go, the student will not feel safe going without knowing if he can return. When we teach the student departure and return, he gets knowing he will need only a little effort to do the job. And being able to check whether he won the game or not, his interest in playing will be increasing.

Of course we can create games to reinforce mathematical content.

Every game, moreover, involves Mathematics, to a greater or lesser degree. So a purely mathematical game is a game par excellence. And if there are games called "games of chance", in the sense of bad or good luck, and if people are dependent on them, Mathematics has no responsibility for that, just as the Atomistic is not to blame for the explosion of the nuclear bomb, which is in the account of human weakness. If an individual uses the light of the sun to blind himself, the sun is not to blame.

Later on we will present examples of arithmetic games, which can be used to make the contents better assimilated, without losing sight of the fact that an exercise in Mathematics itself is already a game.

Steps. The key is not unique. What served Little Paul, who was the realization that he only needed to know how to take the real proof of the division, can serve for very few students who have suddenly faced a barrier in the study of Arithmetic. Paul's father was a merchant and those who have parents of this profession hardly have rejection of numbers, because they experience at home the importance of this instrument. When I saw that the lock on his way was just learning the real proof, I unlocked the door and he moved on, resuming his dedicated student routine.

Learning the ten digits, and then the mechanism of positional notation, to form the numbers, following from there to the four basic operations, are relatively simple steps and only in excessively pathological cases a child presents learning problems in those early stages. The problem can start in learning the multiplication algorithm, for some, or division, for a larger amount. But then, the number of these kids with learning problems is small.

There is an anecdote about it, in which the person complains: "I was doing well in Mathematics, but from one moment to another they added the letters in the accounts, and I was lost there." As one can see, it is just a joke. Most children keep getting good grades in arithmetic topics with natural numbers. In the division many already begin to show resistance, but the problem is still not glaring. Prime numbers, divisibility, factoring, power notation, least common multiple (LCM), greatest common divisor (GCD, or HCF: highest common factor), all these subjects are very palatable to most.

The first knot occurs in the addition of fractions with different denominators.

At that moment the student, who is called Andrew, has left already the operations with non-negative integers and has begun the study of rational numbers. The first approaches with numbers in fraction form are easy to assimilate for children. Mixed number, identification of proper and improper fractions, identification of apparent fractions, equivalent fractions, inverse of a fraction, simplification, irreducible fractions, all these themes are simple and if there is any complaint is by the vastness of concepts and techniques, not by the difficulty in mastering each one of them. They are many, but they cannot be lacking, because they are the basis for what comes next. When the student begins the study of the addition of fractions with equal denominators, he sees that it is a very easy point, but he begins to suspect that there is the preamble of something not very banal.

From one moment to another the professor leaves behind the fractions with equal denominators and shows the addition of fractions with different denominators. The student Andrew cannot repeat the denominator and then add the numerators, as he had been doing before, because now there is no common denominator. For the student, it is not the moment of the exchange of the digits for letters, as it is said in the anecdote, but the moment of the exchange of the equal denominators by the distinct ones, in the operation of addition, that scares and frighten the majority. We are talking about the country with that Iberian heritage commented at the beginning of this text.

If we are in a country like China, Finland, Canada or Italy, students know that they will have to face a more laborious topic, not a more difficult topic. It will be a matter of dedication, with attention and resolution of exercises, and the stage will be fulfilled. In Latin American countries, as in the United States, whose greater part belonged to Mexico, most students will see the new subject as something that has smell of an insurmountable barrier. It is none of this, but the sensation comes from ancestral fears and it overcomes the capacity for sound perception of reality.

For the first time the child is faced with a multi-step math operation, which he has to assimilate, quite unlike the problems with one or two steps he was accustomed to. Let us look at an earlier situation, which for some is complicated, but for most it is not. Let us take the problem of finding the least common multiple of the numbers 8, 15 and 20. By the method of simultaneous decomposition in prime factors, the first step is to assemble the resolution scheme. The student writes the three numbers, separating them with a comma (8, 15, 20), and on the right he traces a vertical line. There, the account was armed, or assembled. The second step is to remember the sequence of the first prime numbers, from 2 onwards, and to effectuate the division one by one. He writes 2 to the right of the vertical line, ahead of the three numbers, and puts under each of them the result of the division by that prime number. Under number 8 he puts 4, under 15 he puts the own 15 and under 20 he puts 10. That 15 was repeated because he learned that when the number is not divisible, in this method of simultaneous decomposition, it repeats itself on the bottom line. The values will now be 4, 15, and 10. He again writes 2 to the right of the vertical line because there are still values divisible by 2. And this is repeated one more time because it will have line 2, 15 and 5. In the line below of this, there will be the numbers 1, 15 and 5. He goes to the next prime number of the sequence, which is 3 (it is not yet time of 5, despite the temptation). The numbers 1, 5 and 5 will come in the following line. Finally, only this class will remain divided by 5, the next prime number. He will have line 1, 1, 1. The second step has been completed. To have the least common multiple, it is enough now that he multiplies the prime numbers he got right from top to bottom: 2, 2, 2, 3, 5. The product gives the answer, 120.

If he wants to sophisticate, he can write the product in the factored form, which is 2^3*5*5, but this is not necessary now, once what was asked for was the value of the multiple. In practice, there are two steps, as we have seen. The first, the assembly. The second, the divisions, by the first primes, with the results under the line of the three numbers. Once the account is over, the student has the factors arranged vertically, to the right. Multiplying the five values could be considered as a third step, but it is a phase that is already won if the student has correctly fulfilled the second step. Anyway, the first step, which is to arm the scheme, and that third, which is the multiplication of the resulting prime numbers, are too easy.

But the teacher did not ask the student in advance to calculate that multiple. He asked, yes, to add the fractions 3/8, 4/15 and 7/20. The student Andrew was accustomed to copying the denominator, which was equal in the given plots, and to add the numerators. Now everything has changed, or almost everything. If he paid attention to the explanation, or if he has an example to follow, he will see that after writing the three fractions with the "+" symbol between them, 3/8 + 4/15 + 7/20, he will have to "find out" that equal denominator, or common denominator, which will allow him to account for the way he has been doing before, by writing three fractions with the same denominator. In the copybook, Andrew writes 3/8 with 3 above and 8 below the fraction bar.

Let us count the steps. The first was to arm the account, i. e., to write the fractions with the sum symbol between them. The second will be to write after the equality symbol, "=", three fraction bars indicating fractions to be summed, with a "=" at the end: --- + --- + --- =. So far, any student who does not know any account can go. The third step will be to discover the common denominator, to write it repeatedly under the three bars. The student who learned the method of simultaneous decomposition calculates the least common multiple, according to the previous paragraph, obtaining the value 120. If we take into account that there were three steps, and there were already two now in the addition of the fractions, the student now enters the sixth step, which is the writing of the equal denominators with the consequent obtaining of the three corresponding numerators. Since he has altered the three distinct old denominators by three equal denominators, writing three times the number 120, he must find numerators that form with this denominator fractions equivalent to the three original fractions. To find out the factors, he uses the resource of the multiplication by the inverse number, which is the division. If the first denominator was 8 and now will be 120, the factor used to go from 8 to 120 is obtained by dividing 120 by 8, and this factor applied to the numerator will give the sought equivalent fraction. He will divide 120 by 8 (the value below) and multiply that quotient by 3 (the value from above). He will do: 120 divided by 8, which gives 15, and 15 times 3, which gives 45. This is the value of the first numerator. Now he will do 120 divided by 15, which gives 8, and this 8 times 4, getting 32. Finally, 120 divided by 20, which gives 6, and this 6 times 7, which gives 42. Ready. The sixth stage is completed. The seventh step is to do the operation he already knew before, which is to add up the three new numerators and copy the common denominator, 120. It will have 45/120 + 32/120 + 42/120, which will give 119/120.

There is still an eighth stage to be accomplished: Checking if the fraction is irreducible and, if it is not, simplifying it. The "computational" way of doing this is to get the greatest common divisor between numerator and denominator. One divides up and down by this divisor and one has the final fraction. The intuitive way is to traverse all the prime factors from the bottom up. The student verifies that 120 is divisible by 2, but the numerator is not, because it is odd. Then he verifies that 120 is divisible by 3 (the sum of the digits 1+2+0 is 3, which is a multiple of 3). He tests the numerator by making 1+1+9. This gives 11, which is not a multiple of 3. He checks divisor 5, looking at whether the terms end at 0 or 5. The denominator ends at 0, then it is divisible by 5, but the numerator ends at 9. Nothing is done. Prime numbers from 5, like 7, 11, 13 and following ones, do not need to be tested, because the largest prime number that came up in the factorization of 120 in the account he made, of the third step, was number 5. The fraction is irreducible and the eighth stage is accomplished.

If the fraction was not irreducible, he would have simplified it by dividing numerator and denominator by one of those prime factors, assembling the final fraction.

If Andrew did not know how to find the least common multiple, but knew how to simplify fractions very well, he could circumvent the situation by multiplying all denominators and using a multiple almost always greater than that he would have found as least. If he has achieved the result correctly, the teacher should not punish him for having exchanged the least multiple for a multiple of convenience. The ideal thing, however, is that the learner has learned the previous topics, or that he takes advantage of the moment of the demand for them to finally learn them.

3. The passage

Either way, you can see, reader, the rite of passage that was set up at this point in the student's school life. From small problems of one or two steps, rarely three, he now faces exercises involving eight steps!

Does it make sense to say "never mind" and let the student move forward without learning to add fractions? Should parents and teachers turn a blind eye at this? The answer is the administrator's classic output: "It depends."

To receive promotion at the end of the school year the student does not have to obtain grade 10, theoretically showing that he has mastered 100% of the studied content. He has to score 5 in almost all school systems. So, knowing other points, but ignoring addition of fractions, he can move on, but he should not be allowed to give up that topic. As Mathematics subjects are almost always recurring, he will have new opportunities to retake the theme. If the school does not offer him this opportunity, it is a bad school.

He does not need deeply to learn the subject the first time he studies it, but he must learn it later. Does not he want an exact science career? Does not matter! The study of fractions is not only for those who are going to study Engineering, Medicine or Psychology. It is for all human beings!

If the student did not learn to manipulate Roman numerals, to write 953 (CMLIII) in this notation, for example, this is an important item for history and general culture. He will not know how to read dates on ancient monuments, what can be humiliating factor. But for the continuation of mathematical studies, the relevance of that topic is very small. Thus, there are some subjects that are limited to some stage of school life and have little repercussion in the future. If the student did not learn the names of the capitals of the Balkan countries, he can hide this deficiency from others for the rest of his life without losing anything of it. He can even win a Nobel Prize in some of the areas covered by the Swedish Academy. It is not the case of adding fractions. If the student is not able to incorporate into his apprenticeship that eight-step basic technique for adding fractions, he will hardly accomplish other tasks involving several steps.

- I disagree, my refractory questioner says.
- Wait, I have not finished the reasoning.

The one who said that he disagrees is based on personal experiences, for having met people who do not know how to add fractions and who, even so, perform well in their fields of activity. Yes, the citizen cannot be a

Arithmophobia: How to Heal the Horror of Mathematics

good aerospace pilot, or a good computer programmer who deals with encryption problems in banking systems, but he may have good acting in fields that do not require a good knowledge of Arithmetic and Algebra. Why does this occur? Because we live in a society with a high proportion of victims of arithmophobia. And these victims need to earn a living. They must show good performance in the trades of their choice.

- I'm a prosecutor and I've never had to add fractions in my work, so if I had learned that, I would have wasted time for nothing, one of the victims of the disorder says.

- No one wastes his time to learn any basic math topics, I say. You are a good prosecutor, but if you had learned to add fractions, you would be a much better prosecutor today. This holds true for any profession.

- I still disagree with you, he says.

- I know, I respond to him. Many victims of arithmophobia have the problem as an ingrained evil and no longer think about curing it, this being your case. A chemical engineer who suffers from melophobia will always say that the absence of music causes no inconvenience in his life, but without knowing Bach, Mozart, Fauré and Verdi, he will never know what he is missing.

A young person should not cultivate arithmophobia throughout his life, leaving the problem to become a pet handicap. The sooner we learn a subject, the easier it will be - if we master the prerequisites for that learning. And the sooner we try to deal with a psychic transformation, the better conditions we will have to escape quickly.

Importance. The addition of fractions is therefore the passage of the phase of the tricycle ride, with safe tasks and a few steps, for the stage of the more exciting bike ride.

A fifth-grade elementary teacher told me that the topic of Arithmetic in which students had the most difficulty for assimilating was that of decimal operations. I wondered, and asked if it was not fractions. She said no, and went on to say that the difficulty was the decimal numerals, which students could not really learn. I spent a few months thinking about it, trying to understand how students who learn fractional operations can show difficulties in operating with numerals with a decimal dot, if an only class for addition and subtraction is enough for them to learn both operations, and then a multiplication class and another two or three of division for the matter to be exhausted. The subject fractions, with its preambles, from what we have seen above, demand months of study. How to understand the teacher's statement? Now, when I answered that the greater difficulty was in the fractions, and she disagreed, reaffirming the position of before, I did not ask whether she in fact taught the little ones the subject fractions.

The only explanation for someone to find that the most difficult topic of learning in the elementary course is that of decimal numerals is this: fractions have not been taught before. Now, for a student who learned fractions, to do the operations with decimals later is like to stop cycling and go back to tricycle riding.

The emergence of fractions occurred in the thirteenth century, at least in Europe, when Leonardo of Pisa, the Fibonacci, taken by his father, the merchant Bonaccio, learned this new way of dealing with numbers together with Arab traders who traded in the Italian Peninsula. The boy, who later became the greatest European mathematician of the Middle Ages, disseminated the use of fractions in European culture two millennia after Pythagoras consecrated whole numbers as the last jewel of Mathematics, along with Geometry and Music - he also included Astronomy, but this was more object of observation than of recurring conclusions.

Decimal. Decimal numerals are a much more recent subject. They have been created in the seventeenth century, by John Napier, in Scotland. They are a new method for the representation of rational numbers, which until then, discounting some incipient attempts, were only represented with fractions. Learning decimal numerals without knowing fractions is a task for Hercules, if Hercules once dedicated himself to numbers, besides counting from one to 12.

The subject had been developed by other scholars, in a preliminary way, and it was up to Napier to propose the final form of the notation that came to us. Predictably, he took into account the difference in the tradition of writing numbers in Britain and on the continent, and proposed that the point was used among Anglophones, and that the Europeans of the continent should use the comma to separate the decimal part of the integer part. We must remember that Anglophones already used the comma to separate the thousands in the number, as in 35,452,000, which is an integer, while the continental ones used the dot. As his goal was to use notation in logarithms, an invention of him, this broken part of the value of the decimal logarithms has been called "mantissa", taken in its positive value, nomenclature that remains to this day.

In March 2008, the US Department of Education in Washington, DC, released the "Final Report of the National Mathematics Advisory Panel", presenting the diagnosis made by math teachers in the whole country. The conclusion was that the problem was in the low learning of fractions, measures and Geometry. Then it came the recommendation that education systems should reinforce these issues. When one moves with measures, practical sense recommends to use decimal numerals, which are more popular than fractions. If we go to the store to buy 1 meter plus 3/4 of fabric, we do not ask the seller for the fraction value, but we say 1.75m. That is why the school, which prepares for science rather than for the market, does not ignore the teaching of decimals already in primary school. In fact, currency units, with their pennies, are treated with the decimal numerals.

Pocket calculators, in the same spirit, operate with decimal values. That is what the marketer and the newsboy need. A calculator working only with fractions would serve very well, but only to students and researchers, not to the larger clientele, consisting of merchants, industrialists, and workers in general.

The student who imagines being liberated from learning accounts because the calculator already provides the necessary results is living in terrible deception. Firstly, to be clear about how decimal numerals work, it is necessary to know the fractions well. Secondly, Blaise Pascal invented the calculator, in 1642, to relieve the work of his father, Etienne Pascal, who was a tax collector in France. Seeing his father spend dawns sided by a candle amid stacks of papers with additions and subtractions, young Blaise invented a calculator of crank. Turning it clockwise, the sum was made. A few years later he added to the "Pascaline" (the machine nickname) the ability to do subtraction, turning the crank backwards. Some decades later, Leibniz, in Germany, added the multiplication and division functions to the machine. The machine was invented, therefore, to aid in the work, not in the learning in the classroom. The third point is the one that most justifies the learning of fractions, and it deals with numerical precision. When we need to use the value of the fraction 20/3, without loss or gain, we leave this number in fractional form and the value is all there. If we translate it to decimals, to use it in the calculator, for example, we see that the division falls into a repeating decimal and we have to decide the precision we want. If it is hundredths (it can be cents, for example), we write as 6.67, rounding up the second house after the comma, according to the universal rule. As the division gives 6.6666... and we want only until the hundredths, the rule says that by ignoring numbers between 5 and 9, including these, we add +1 to the last house that stays. If the number is from 0 to 4, we keep the house as it was. For example, 9.52333... is 9.52, again taking into account that we want the hundredths house. If we wanted the house of thousandths, we would have 9.523.

Now, when we enter 6.67 instead of 20/3, we make a change from a hundredth up. If we multiply this by a million, the discrepancy will not be now of only hundredths. We see that the decimal numerals are convenient for day to day, but not for science in general. We will always be in doubt whether what we do in Mathematics is true or not, as Bertrand Russell said, but if we can choose between something intrinsically precise and something that depended on rounding, of course we are left with what guarantees us precision.

Faced with the above three reasons for the importance of learning the operations manually and mentally, without relying on calculators, the teacher has a powerful ally in the classroom, which is getting the rest of the division. A cheat student can use the calculator in secret to get, for example, the quotient of the division between 20758 and 95, whose entire part is 218. Now, if the teacher asked him to find the rest of this division, with only a lot of juggling he will have it using the calculator. If he has any skill similar to that of Pascal or Gauss, he will know that to find the rest it is enough to apply the Euclidean formula of the division, by multiplying the quotient by the divisor and subtracting that product from the dividend. The rest is then obtained by making 20758-218*95. The majority's way, however, is to make the complete division and obtain the value under the dividend, and in this case this remainder gives 333. The one who depends on the calculator and deceives the teacher, only for a miracle he will be able to use Euclid's formula to find the rest. If he, untrained in this by another, deceives the teacher and yet has the insight to apply Euclid's formula on his own ($D = q*d + r$, dividend equals the quotient times divisor product, plus the rest), then the school unit has among its students a genius of dishonesty. The teacher implicitly taught the formula, since it is used when taking the real proof of the division, but the application in the specific situation of finding the rest is not a common step for the child.

Before proceeding, I use the above statement to make an alert. The learning of Mathematics is a universe in itself, almost always simple and harmonious, whereas the apprenticeship of the application is a second universe, this yes, marshy. Those who want to make Mathematics a scrambled ball of yarn go to this job of requiring students to make applications for which they have not yet matured. Pythagoras, Plato, Euclid, Galileo and Poincaré asked their students to solve problems of Geometry. Confused teachers ask for solving Electrical Engineering problems that their students have never seen before.

While discounting the case of the victims of advanced arithmophobia, is it possible that someone who can write a four paragraph message without grammatical errors is unable to learn addition of fractions? No. Unless some brain dysfunction, unidentified to this day, is discovered preventing an articulate and intellectually functional person from learning certain numerical processes. It is not impossible, as there may be an individual with a physical injury in the brain in the region responsible for numerical processing in a situation where this does not interfere with other mental activities. But that would be something as rare as the case of a person who sings the chromatic scale of C major without being able to emit the note G, making C, D, E, F, _, A, B. It would be something completely unusual, but not completely impossible.

Treating people with no physical or physiological oddity, as in the case of the singer above, it is expected that if a person can be articulated by expressing himself in his native tongue, he has no difficulty in learning fractions.

The insistence on the addition operation, which is the same as subtraction, with the difference of the signal, and the subtracting account itself, is because the operation of multiplication of fractions is one of the easiest in the school life of the child in the elementary courses. If we want to multiply two fractions, we just need to multiply numerator by numerator and denominator by denominator. If the result is not irreducible fraction, we must simplify. And as we have to simplify, as we move forward in learning, we tend to simplify the factors before proceeding to the final operation. Those who do not acquire this skill do not have to worry, at least in the numerical phase of their courses, as they can always simplify at the end of the account. After learning multiplication of fractions, obviously division of fractions comes. This is easy. The student has only to remember that division means multiplication by the inverse. So, divide 3/5 by 4/7 means write 3/5 multiplied by 7/4 (inverse of 4/7) and make the count to find the product.

When the student finally completes the items on fractions and moves to the decimal numerals, he becomes aware that a fraction is an indicated division. For example, to transform 4/5 into decimals we divide 4 by 5, obtaining 0.8. Since 4/5 and 0.8 have the same value, it is clear that this 4 is a dividend and that 5 is a divisor, which fades into the new expression 0.8.

Now, the book or the teacher would greatly help the student if it informed us from the beginning of the study of fractions that this form of numerical writing represents an indicated division. If it is in the form of fraction, division does not have to be made, for a number like 4/5, four parts out of a total of five, already has a meaning in itself. But the meaning is 4 in five, and this is the same as 4 divided by 5. In the fraction 1/2 the fact becomes clearer, because taking 1 in 2 is taking half, indicating that the two have been divided in half. School textbooks that give the student this information are very rare. And it does not cost more than a leaf line.

In addition to all that has already been said about the importance of learning the topic fractions, complemented by decimal numerals, one cannot be overlooked: that by learning them the student is ready to learn all the arithmetic and algebraic themes that he will have to face. Nothing else will be mysterious. And is there anything more comfortable than this? Of course, subjects are changing and new concepts will come, so the student will not get high marks if he does not study and practice. But he will have the confidence that, having learned fractions, nothing else will frighten him in Mathematics.

Challenge. There is only one setback in this categorical statement that I make. When the teacher receives a new class, a ninth-grade one, for example, he does not know whether the students have had all the previous topics in junior high school, whether they have been left blank, lacking a teacher, or having a teacher slow. But when the class comes with the same teacher since the seventh grade, then the teacher knows that what is required and is not fulfilled, due to lack of knowledge, is related to difficulty of memorization or individual ability. I has had the opportunity to go through a challenge that intrigued me a lot. In a ninth-grade class who had been in Math with me since the sixth grade, by putting questions of irrational equations on the test, I saw that nobody was able to solve its most complete type, which is the one in which the student has to eliminate the radicals more than once. The simplest case is that in which the apprentice raises to the square the radical that he isolated in one of the sides, also raising the other side, eliminates that radical and continues the equation with the remaining variables and numbers. In the more laborious case, there is more than one radical, among other plots. One of the radicals is isolated in one side, but in the other side of the equation there are radical and more terms - if there is only one radical left in each member, raising everything to square eliminates it and everything becomes easy. By raising the two sides squarely, the side that has more than the radical remains with radical. If I put in parentheses a root of x, plus 3, and raise this expression squarely, I have to apply the first special binomial product, and get x, plus 6 times root x, plus 9, that is, the radical continued. The student must apply a recursive reasoning and resume the work of isolating the radical and repeating the procedure, now eliminating the nominee. My kids could not do that.

Arithmophobia: How to Heal the Horror of Mathematics

As the failure was of the whole class, although they had gotten good marks from the resolution of the other questions, I warned that I would continue to put that kind of question in the next tests, until they got it right. I do not remember if I did it two, three or four times. In one of these editions, I corrected a test and saw that the student hit everything. His name is Fabio. He is a nisei, son of a civil engineer. I corrected the other tests and saw that nobody else was right. Maybe I was wrong on the level of demand because this class was strong and got blue grades. If I had tightened the siege, they might have given me a little more. Probably the student Fabio was the only one to take seriously the challenge of studying that type of exercise. Either way, I was relieved to see that students can solve complicated problems for their age. Fabio solved it and I decided not to insist more with the others. If I had insisted and increased the level of demand, more students would have been right.

The fact has brought me to the awareness that problems that require several steps in their resolution are the great knot in the students' lives. If these steps require recursion, as in the case of having to square the side more than once in the equation, then the degree of difficulty becomes more acute.

Firmly holding the rudder and going through the storm without fear is not for a first-time sailor. The training, the accumulated learning with the training, is growing and giving confidence to the student.

Doing math exercises is much safer than any other scientific activity. The student does not get wet, does not carry weight, does not get sweaty, does not get pain in the body, does not receive an electron discharge in the face.

In addition to practicing the science on which all others depend, and on which depends the progress of mankind, every exercise he does can be seen as a game, which he will gain by coming to the answer.

When I write a book or compose a song, I do not know what result I am going to get, exactly. They are open works, as Umberto Eco said. When I solve a mathematical problem, it has a result, which I must discover. It is much more playful.

Practicality. For the child to feel more excited, whenever possible a practical problem should be offered. If the problem is geometric, it is already visible, or viewable, and that helps a lot, but a problem of trade, travel, distance or time shows how Mathematics can be used in daily life and gives a greater sense to learning. What one cannot do is give the impression that the heart of the matter lies in practical problems, because this is a false idea. A practical application should have the role of reinforcing learning, not confusing the meaning of the question

As was said above, the practical application is a second universe, which transcends Mathematics, and the teacher should never forget this fact.

When the student performs an operation with numbers, he is doing a general, or, if he prefers, generic, operation. When he takes values out of a practical problem and does the math to get to the solution, he is dealing with a specific situation. That is why practical problems must be done as a complement to learning, i. e., as broadening horizons. It is clear that a practical situation can be used as a motivating element to arrive at general concepts. But the ideal is to learn the general situation and apply it in a specific situation, not the contrary. If the examiner wants something closer to the concrete, he can use geometric situations. Instead of a piece of wood, he can speak of a segment. Instead of a pizza disk, a circle, simply. Later on, he uses the pizza platter, with all the appetite it can open.

Here is an example of a practical situation. Paul had to walk a certain road in three days. The first day she walked 1/3 of it. On the second day he walked another 3/5. Which fraction is left to go on the third day?

First our student Andrew will add the two fractions, obtaining 14/15. To find out what is missing for the entire road, which is the integer 1, or the fraction 1/1, he will do the subtraction 1/1 - 14/15. Applying the common multiple 15, he will have 15/15 - 14/15, and will reach the result 1/15.

For the last subtraction operation, Andrew can take the real proof, adding the subtrahend 14/15 with the result 1/15. He sees that he gets 15/15, which, simplified, gives the integer 1, which is the whole road.

Now let us look at a purely theoretical problem, but based on Geometry. The student Andrew drew a square, with a height of approximately 4 centimeters, in his notebook. With the pencil, he divided it in half horizontally and vertically, forming four congruent squares ("congruent": of equal measures). Then he painted the inside of the square in the upper right corner with his pencil. Then he divided the square in the lower left corner, from top to bottom, into two small rectangles. Then, of the two rectangles formed, he painted the one on the right. One asks for the fraction that represents the painted area inside the original square.

To solve, Andrew notes that by painting the square in the upper right corner, he painted 1/4 of the full figure. By dividing the lower left corner and painting the right rectangle in half, he painted 1/8 of the large square. He knew it was 1/8 because the bottom corner is a square representing also 1/4 of the total, and he divided that square in half. He knew half a quarter gave 1/8, but he wanted to check it out by doing the math. Knowing that a fraction "of" another fraction is the multiplication of the two (that "of" between fractions means "times"), he effected 1/2 times 1/4, obtaining 1/8. Then to find the whole painted area he added 1/2 with 1/8. Applying the common multiple 8 and making the sum of the two resulting fractions, 4/8 and 1/8, he arrived at the answer, 5/8.

In order for the student to learn fractions, ass well any other rich topic of Mathematics, it is not silence that he needs, but respect, dialogue, attention and dedication. When the teacher is explaining the subject, the silence is broken by his speech. The next moment, of the exercises, the students need to talk about the topic in question. There must be silence, yes, during individual evaluations, because at that moment communication must be prohibited, once student fraud is one of the greatest enemies of learning. Even the old custom that some teachers have, to explain the test questions before students try to decipher them, is a damaging practice, because it delays the development of the ability to interpret.

Percentages. As Andrew also studied the operations with decimal numerals and the percentage calculations, the teacher asked for a part B in the resolution of that exercise. The previous one was part A. He asked how much percent of the original square that painted area represented. As Andrew had already obtained the fraction 5/8, he only had to pass this result to the notation of decimal numerals and write the corresponding percentage. He divided 5 by 8, in rational operation, that is, with a dot, and got the value 0.625, which reads as "zero dot six two five" or as 625 thousandth - if the division of 5 by 8 were in integer operation, he would have obtained 0 as quotient, which, multiplied by the divisor and subtracted in the dividend, would give rest 5. As he wants to know how much percent this value is representing, and "percent" means "hundredth", then he wrote 625 thousandths as 62.5 hundredths, which is the same as 62.5%. In practice, Andrew already knows that any number written with a dot is transformed into "percent" by moving the dot two houses on the right. From 0.625, just make the dot jump the "6" and the "2" and write the "percent" symbol. What he did, implicitly, was multiplying 0.625 by 100 and adding the symbol "%", which represents multiplication by 1/100.

Before he came up with this ability to move with percentages, Andrew had learned the four operations with decimal numbers very well.

For addition, he learned that the account is armed by arranging dot under dot. For subtraction, besides to assemble the numbers arranging dot under dot, the spaces on the right should be filled in with zeros to complete the decimal place along with the number that has more digits to the right of the dot. For example, to make 23.45 minus 8.7592, at the time of arming we write 23.4500, because the number that will come below this, 8.7592, has four decimal places. To write it in the account we put the figure 8 under the figure 3 of the minuend, and thus the dot of this number will be under the dot of the other.

To multiply numbers in decimal notation, it is irrelevant to write a dot under a dot, because the dot position in the result comes from another type of observation. Just count how many houses after the dot are in the first number and add up with the number of houses after the dot in the second number, and this will be the number of houses after the dot in the result. For example, to make 12.3 multiplied by 7.915, we multiply these factors as if they were integers, without concern for dots. After completing the multiplication of integers, we count the houses after the dot in each one. In the first, there is only one house. In the second, there are three houses. The total is four houses after the dot and this is what should be observed in the final result. Before entering the dot, the number that appears is 973545. As we count four houses, we will be with 97.3515.

Finally, Andrew learned that to divide numbers with dots, the most convenient is to equalize the number of houses after the dot, as in subtraction, and to bring the dot to the end, both the dividend and the divisor. Thus, for 72.4 divided by 1.521, we wrote under the division bar 72.400 and before it, 1.521. Since there are three decimal places in each, we eliminate the dot, dividing now 72400 by 1521. He already knew this in the case of a division of 5 by 4, by lowering the last number that creates a value greater than or equal to the divisor, next to be downloaded, even if it is zero, we put a dot in the quotient before writing the resulting digit. In 5 divided by 4, the first step gives 1 in the quotient, with 1 in the rest. There will be no more digits to go down in the integer part, meaning that the 5 represented the last of that nature. The next one to be downloaded, since no other appears after the dot, is 0. So before writing the result of 10 (1 plus the digit 0, downloaded) divided by 4, which will be 2, we write a dot, which will be between the 1 and the 2. The final result will be 1.25.

4. Basic operations

When we speak about elementary arithmetic operations, even including operations with fractions and decimal numerals, we understand that they are four: addition, subtraction, multiplication, and division. Now, to find potencies and roots may not be very elementary actions, but we can call them basic arithmetic operations, since they are learned in the primary and junior high school.

Are there six elementary arithmetic operations? The writer of scientific dissemination Yakov Perelman swore that there are seven. The seventh operation, he wrote in the book Algebra For Fun, is the calculation of logarithms, the invention of John Napier of which we have already spoken.

Yakov Isidorovitch Perelman was born in 1882 in the Russian city of Grodno, which today belongs to Poland, and died in 1942 in St. Petersburg. Graduated from the Institute of Forestry Technology of this city, he gained renown as a writer of science fiction and popularization books on themes of Physics and Mathematics. He died of starvation, two months after his wife Anna Davidovna had passed through the same fate, during the Nazi siege to the city.

The Russians developed a strong attachment to the exact sciences in the first decades of the twentieth century, and one of those responsible for this is the writer Perelman. Some of his important books are translated into Spanish, Portuguese and English and are a powerful tool in the hands of families who want to take children out of the dread of Mathematics.

Logarithms. In the case of the classification of the logarithms as the seventh elementary operation, he may have exaggerated a little.

His justification is the one that comes next. The four initial basic operations, each operating as the reverse operation of the other, form two pairs. Subtraction is the inverse of addition and division is the inverse of multiplication. When we consider power, we take the inverse operation of the root, which results in the base of power. For example, 3^2 gives 9, and if someone gives me the number 9 to find the bases of the power that generated it, the inverse operation will be the square root of 9, which we know that is 3. In this pair of operations it was implied that the exponent of the power was 2 and this same 2 was used as index of the root: we could not have done cubic root, of index 3, but square root. Now when we extract the square root, root of index 2, of our power, we find the base. When we apply the logarithm calculation in this same power 9, knowing that the base is 3, we find the exponent. In this work of powering we have not only power and root, but also logarithm. When we apply logarithm to a given power, with base **b**, we discover the power that on that base generated the power, or logarithming, that we have. Thus, in addition to the four basic operations of the primary course, we now have three more operations. The total is seven.

It also weighs in favor of Perelman's position that in previous operations the reverse is equally applicable to any of the components of the starting operation. For example, at 4+5 = 9, we can apply the inverse by doing 9-5, which gives 4, or 9-4, which gives 5. If we multiply 4*2 = 8, we can apply the inverse by doing 8:2, which gives 4, or 8:4, which gives 2. In the case of power, differently, the operation changes. In 3^2, to find the base we use the root, while to find the exponent, we use the logarithm.

And why is not accepting this fact so automatic? Why do not we institute now that the calculation of logarithm is the seventh elementary arithmetic operation? The reason is the calculation process itself. It is possible to find logarithms using only arithmetic, but the work with the operation is almost always presented with the use of algebraic resource. To find the value of a logarithm on a given basis, we equate this logarithm with a variable **y** and, by applying the inverse operation, which is the power, we find the value of **y**, that is, of the involved exponent.

To find the logarithm of 9 in base 3, $\log_3 9$, we equate the expression to **y** and, by the inverse operation, we write a base power 3, with that exponent **y** (remember that **y** is the logarithm and this represents the exponent) and we equate to logarithming 9. Solving the exponential equation $3^y = 9$, we find the exponent (just factor 9, making $3^y = 3^2$ and cut 3 with 3).

By the purely arithmetic method, we apply the third property of logarithms, which ensures that the logarithm of a power is equal to the exponent multiplied by the logarithm of the base. Thus, in $\log_3 9$, which is $\log_3 3^2$, we "take" the exponent 2, arriving at $2*\log_3 3$. As logarithm of the own base is always 1 (the base 3 raised to the exponent 1 gives 3), simply we multiply 2*1 and we will have the value of the logarithm, 2. (The four properties are: first, logarithm of product, $\log_a(u*v) = \log_a u + \log_a v$; second, logarithm of division, $\log_a(u/v) = \log_a u - \log_a v$; third, logarithm of power, $\log_a x^p = p*\log_a x$; fourth, change of base, $\log_b x = \log_a x / \log_a b$.)

Everything seems very simple, but the assurance that the logarithm of the base is always 1, as well as that of the logarithm of power is the exponent multiplied by the logarithm of the base, this guarantee is obtained by algebraic process.

Logarithm, despite its strong arithmetic component, is an algebraic theme.

A lady I met many years ago, who had only completed primary school, once asked me:

- Does school still teach logarithms?
- Of course, it teaches, yes, but how do you know about logarithms, if you only studied until fourth grade?
- I studied the subject in fourth grade, yes, she replied.

The fourth primary grade before is the fifth grade now in Brazil, for 10-year-old students. Yeah. In the fifth grade, completing the arithmetic operations, school taught logarithms. But I may have made a mess, when she told me that she went through fourth grade. Maybe she was referring to the fourth junior high school, which is the ninth grade today. Anyway, it was too early. Certainly it was so in the city of Perelman too. Over time, those responsible for the distribution of topics in the curriculum were realizing that the subject should be addressed later, within the field of functions, with algebraic treatment, in High School.

Facilitation. Students will always complain about the number of topics they have to study and the degree of difficulty of the subjects, but the simplification in the last decades has been great. A few years before I graduated from High School, students in the "Scientific School" mode studied in Mathematics the chapter Series, a vast and profound subject that belongs today to the second year of university. In Physics, still in my time we were studying gravity problems with a value of 9.78 m/s^2 for acceleration, not the rounded value of 10 m/s^2 of today. In the chapter on Electricity, we studied the mesh of currents and tensions using determinants, to apply Kirchhoff's Law. The units, already in the first year of High School, did not follow the unification of the International System, but were interchanged, sometimes in the MKS (meter, kilo, second) system, sometimes in the CGS (centimeter, gram, second) system. These are just a few examples of how the accounts were more laborious in the twentieth century compared to the twenty-first century. And it was no use for the Physics teacher to allow use of calculators, because they were still crank-based. Liquid crystal display calculators were launched in 1972, using the invention of Pierre-Gilles de Gennes, the great French scientist, Nobel Prize, who died in 2007.

In the final years of the middle course, it is necessary to use calculators in laboratories of Physics and Chemistry, but the school should only admit use of calculator if it is scientific calculator. When the student asks to use a calculator and appears with a fair machine (without demerit), with all four operations and a square root symbol, plus a percentage symbol that induces a brutal Logic error (I will come back to that later), he is a lazy or misguided student, who wants to avoid doing accounts on paper. The scientific calculator, however, as well as the financial one, allows to do operations that cannot be done manually. Almost always, this revolves around powers. In the financial calculator, for example, we can find the root of a 25th degree polynomial. A Physics teacher who, from the first year of High School, leaves students comfortable to use calculator of the four operations in the tests, saying that what interests him as an examiner is the formulas and the concepts, not the Arithmetic, this teacher is working against his colleague of Mathematics. The students of this phase, 15 years-old, are still immature in the use of numbers, so much that they want to make multiples of integers in the calculator. They need teachers to require their training in this field, and the math teacher alone cannot. Many want to create a High School curriculum in which the student has Mathematics and nothing more in exact sciences. It is an anti-educational idea, because Mathematics is consistently learned when the student applies it in the subjects around. The applications are very varied and it is not for the Mathematics teacher to show them, except in one or the other example.

System. Thus, it is when the student needs to work with powers of base 10, or, for example, cubic roots and quintic roots of any numbers, that he must be allowed to use the machines, and this almost always occurs in the laboratory. And to know how to work with roots, one must first master the meaning of the powers.

The idea of base 10 powers lies in the construction of numbering systems. We have used the decimal base for centuries, but in antiquity the base 60 was widely used in Mesopotamia. When the Europeans arrived in America, the Incas used bases 8 and 20, while the Mayans used 20 and 180. The inspiration for using the number 20 must be in our quantity of fingers. The number 180 was approximation for half days number of the year. As for the number 8, it is not known where they got the idea.

Any natural number in base 10, or decimal base, can be written in the polynomial form, that is, in notation of base 10 power. How do we write the number 245, for example? We make: $245 = 2*10^2 + 4*10 + 5$ (this second 10 is raised to exponent 1, which does not need to be explained, and unit 5 is multiplied by the power 10 raised to 0, which, does not need to appear as well, because it is the neutral element of multiplication).

Arithmophobia: How to Heal the Horror of Mathematics

At each position, in this numbering system, from the end to the beginning of the number, we are multiplying the digit by the power of 10 whose exponent is that inverted position minus 1. If I am in the last position, in case of figure 5 above, I consider it position 1, and raise 10 to 0. If I am in position 2, from end to end, I raise 10 to 1, and so on. If the number starts with 4 and has six figures, the polynomial form will start with 4 multiplied by 10 raised to 5, which is 6-1, since the last exponent is 0, from 1-1.

If the system is on the binary base, which is the base used in computers and only uses the digits 0 and 1, the bits, a number in the polynomial form can come written as $1*2^4 + 0*2^3 + 1*2^2 + 1*2^0$ (in base 10, the last digit is 9 and the base of the power is 10, while in the binary base the last digit is 1 and the base of the power is 2, this already understood as base 10 number). The number written on the binary base takes only the coefficients, which already appear in the order of the positional system: 1011. To find out how much is worth in base 10 that number started with 2^4, we do: $1*16 + 0 + 1*4 + 1*1$, and this gives $16 + 0 + 4 + 1 = 21$. And how much does it give 1110 passed to base 10?

We make: $1*2^3 + 1*2^2 + 1*2^1 + 0*2^0$. This gives $1*8+1*4+1*2+0$, or $8+4+2+0 = 14$.

As for the base 60 system, of the Assyrians and Babylonians, it continues to be used today, despite all the efforts of scientists during the French Revolution to unify the decimal system of measures. That base we use in the division of time and in division of angle in degree. One hour is divided into 60 minutes, which, in time, are divided into 60 seconds each. The same nomenclature is used in degree, which is also divided into 60 minutes, which are each divided into 60 seconds. When we write 10 minutes, we do not know if we are talking about time or angle, but if we use the notation of the International System of Units, it is easy to distinguish. If we are dealing with time, we write 10min. If we speak of angle, we write 10'. To write 5 seconds of time, we make 5s, while 5 seconds of angle is 5". The student should never want to gain time by writing an apostrophe to indicate minutes of time, since the apostrophe is minute of angle. Until about the year 1960, many textbooks called these base 60 units "complex" measures, but it was an abuse of language. There is no relation to the concept of complex there. Only the base is different from our base 10.

5. The fiasco of the "Modern" Mathematics

In the initial three triennials, elementary, primary and junior high school, the great change introduced in 1962, in what is called Modern Mathematics, was gradually diluted and practically evanesced. The importance of this change was in the unification of language, for Arithmetic and Algebra, which gained a rigor of which it resented before. The authors of school books did not understand each other in terms of notations and even concepts. With Modern Mathematics, all language was standardized.

Sets. The error was in the exaggeration given to Set Theory. The curricula replaced important topics and methods by large studies of Sets, without this bringing gains for students in terms of learning basis to tackle more advanced subjects. The impression that change makers seemed to make was that it was more important to know Set Theory than to know the traditional elementary arithmetic operations. So much so that in 1973 it came a book called "Why Johnny Cannot Add", trying to ridicule that intellectual fever

In the book, the author, Morris Kline, sought to diagnose the losses that schools have had in learning because of the new fad. Asked by the teacher about how much it gives 3 times 4, Johnny responds that because of the commutative property of multiplication, the result is the same as 4 times 3. Nor did it go through his head that the answer should be 12. That was because his learning time was taken by intermediate concepts that are very important for professional mathematicians, but which for children served more to confuse. Is it important to know commutative property? Yes, but as a tool for calculations, not as an autonomous subject within matter. It would be as in reading, to give more emphasis to the spelling of the word than to the reading of the word itself. The teacher would write the word Envelope on the board and ask Johnny to read it. He would then say "E-n-v-e-l-o-p-e", letter by letter, without ever being able to pronounce the word. It was a situation equivalent to this that Modern Mathematics was producing in the development of children.

Professor Jean Dieudonné, the most notable name among those who formulated and disseminated this change in teaching, did no pilot experiments and could not predict such a situation. But if a lesson can be drawn from that confusion it is that one should not apply too great a change in one system at a time. There is a risk of compromising the learning of an entire generation.

Rigor. What was necessary to introduce was strict language, which included the use of set language. This did not require students to spend whole semesters studying Set Theory, sacrificing studies that have always been central to their intellectual growth.

And it all happened within the context of the Cold War, in the wake of world politics.

On April 12, 1961, Russians sent a man into space, Yuri Gagarin, who circled Earth on the Vostok spacecraft, flying above the atmosphere at a speed of 28,000 km/h. Prior to this, since the launch of the Sputnik satellite in 1957, the Russians had sent 49 dogs into space, of which 20 died on mission. With the dispatch of Gagarin, scientists from the United States and France responded to the call of their governments for a renewal in the education system, under the observation that Western countries were behind in the development of science. The result was the disaster of the introduction of Modern Mathematics.

Jules-Henri Poincaré, died in 1912, the last scientist to master all the fields of Mathematics, had left a diagnosis, "Set Theory is a serious pathology, from which Mathematics will soon be cured." Before the cure, 50 years after his death it came that decision to throw down on the children's head that "pathology". In Brazil there was some resistance, but it did not prevail. Professor Omar Catunda, participating in France in the meeting that made the great decision, while hearing Dieudonné shouting "Down with Euclid!", muttered, "In my country, at least Euclid!"

The cure only began to be outlined with the release of Professor Morris Kline's 1973 book.

Do not think that the problem has been solved completely. Even today, at an advanced stage of the 21st century, many authors write their textbooks on Mathematics with chapters on Set Theory and with a predilection for obtaining numerical results via counterproductive methods based on sets, to the detriment of practical and efficient methods.

6. Correction in language

The adoption of rigor and unification in language was of great value, but we continue, especially in Brazil, to repeat, in books and in classrooms, inadequate jargon when introducing concepts, definitions and techniques.

Prime. An emblematic case is the definition of prime number. In Mathematics, implied ideas, connotations and double meanings should be avoided. But children are still victims of these imperfections. Thus, elementary mathematics books still define prime numbers as the "number that is divisible by 1 and by itself". It is difficult to write a more false sentence, for the number 1 is "divisible by 1 and by himself", and it was never a prime number.

The use of the word "prime" here is in the Italian, or Latin, sense of "first". He is the "first", or the smallest, of a sequence of multiples. Thus, the number 2 is the first of the sequence of even numbers, whereas 3 is the first of the sequence of triples, or multiples of 3. The number 4 is already part of the sequence of the multiples of 2 and, as shown in the Sieve of Eratosthenes, it is already cut off and cannot start a multiples sequence. Eratosthenes, the most famous of the librarians of Alexandria, mathematician and great inventor, presented the practical method of obtaining the first prime numbers, which is his Sieve, or criterion. We write the positive integers up to a certain value, for example 20. Let us mark the first of each sequence with a ball and cut off the ones that come next. The number 1 is crossed out in advance, because if it were considered prime, all the following, as they are divisible by it, would be cut off and nothing would be left. That is why we start with number 2. We make a circle around it and we go tracing the multiples, that is, we count by 2 and cut: 4, 6, 8, to 20. The first uncut number is the 3, so it is prime and we surround it. Let us count 3 in 3 and cut. Then we do the same with 5, with 7, with 11, etc.

If you, reader, did not remember Eratosthenes' method for identifying prime numbers, there it is, in its entirety.

Natural numbers that are not prime, that is, products of other numbers, are called compound numbers. The Fundamental Theorem of Arithmetic states the following sentence: any natural number greater than 1 is prime or is a product of prime numbers.

But how would a precise definition of the concept be?

An irreplaceable sentence, from the logical point of view, is as it follows: "Prime number is every positive integer that has exactly two distinct divisors."

In the examples and exercises, which are always necessary for the learning of Mathematics, it is made explicit that these numbers are 2, 3, 5, 7, 11 and all of the following with only two divisors: The number 1 and what is being tested.

And how would the customary definition be corrected, taking advantage of its didactic concern?

It is simple. Just write it like this: "Prime number is any natural number greater than 1 that is divisible by 1 and by itself".

Since the student now starts from numbers larger than the unit, he would not face the risk of finding the number 1 to be prime. But the old wrong definition gave him the right to find that.

Someone can imagine that prime number is a very silly thing and does not deserve so much discussion. This person is totally mistaken. The idea is not silly and deserves not only discussion, but large investments in research today.

So that thieves do not discover your bank account password, reader, financial organizations, together with universities, private or official, spend great amount of money in the search with prime numbers. The largest of the prime numbers is built, and then prizes are given to those who get one that surpasses it. The greater the prime used, the greater the security in encryption, not only banking, but also governmental, strategic, academic and even communications through WhatsApp. In 2018, October, the largest known prime is 2 raised to 77,232,917, minus 1.

Another ancient obstacle that is still defended by many is the use of unnecessary interfaces in the learning of concepts and in the use of certain techniques. They are called "crutches" by the more critical people. The complication they present is that they represent one more scheme for the learner to learn, and it is a scheme that can be dispensed, what reduces the amount of facts to be memorized.

Divisor. An example is the old cannon, similar to a "hash game" frame, used for the calculation of the highest common divisor. The student has to memorize in which alveolus he writes the quotient and in which writes the rest in each step. Without such memorization, the process does not continue. Now, he only needed to know that after each division, from the largest to the smallest of the two numbers, the divisor, which is not yet the common one, will be used as a dividend in the next account, which will have the previous remainder as the new divisor. Yes, there is a recursion there, something not very intuitive for children, but it only involves two transformations: Divisor of the previous account becomes dividend and the rest of the previous account becomes divisor. At this point, our student Andrew has already learned that no positive integer is divided by zero. So he knows that when it reaches a rest zero, it cannot be transformed into a divisor, and the last divisor in the bar is the highest between the divisors common to the two starting numbers.

As an example, Andrew shows how he gets the highest common factor (divisor) between 30 and 21. He divides 30 by 21 and sees that the remainder is 9 (the quotient is 1, but it is irrelevant unless it is wrong). He goes to the second account, which will be 21 by 9 (divisor turned dividend and rest turned divisor). Without worrying about the value of the quotient (which must be correct), he sees that the remainder is 3. He goes to the third account, which now has 9 as a dividend and 3 as a divisor. By making the math, he gets rest zero, and cannot divide anymore. The final divisor was 3, which is in the division bar. It is the value of the sought Greatest Common Divisor (GCD).

Does Andrew know another method? Yes, that of the term-to-term decomposition (he knows that the simultaneous decomposition is used for the Least Common Multiple). It does the prime factorization of the number 30, which shows vertically 2, 3 and 5, and separately decomposes the number 21, which shows the prime numbers 3 and 7. He surrounds what is common to the two decompositions: the number 3. If these decompositions were 2*3*3*7 and 2*3*3*3*5, respectively, from numbers 126 and 270, Andrew would surround the two factors 2 in one and other, then the two factors 3 of the first, relating them to the first two factors 3 of the second. The highest common factor (or the GCD) in this case is 18. The third value 3, in the decomposition of 270, does not have a common emulous in the other decomposition, and does not enter the result, which only wants to know of the common factors (divisors).

What does the teacher do with two equally effective methods to achieve the same result? The greatest certainty he must have is that he should not devote the same emphasis to both. In view of this, he must evaluate which of the two methods is most advantageous for the student's school life. Sometimes a more laborious and slow method is preferable, if the resources used in it are indispensable tools in the following stages of learning. In the present case, he has plenty of reason to prefer the method of decomposition in primes. And if you do, you must present the method of successive divisions, that first, as a curiosity, making it clear to the class that his preference is for decomposition. But to that colleague who already tends to prefer the decomposition in prime factors, I remember that my preference rests on the method of successive divisions, because the decomposition the student already does to find the least common multiple.

If there are two methods, should students always learn both? Not always. There are cases where one of the methods is clearly obsolete, or it is about to retire. We should not spend time with it. The inconvenience of showing only a method when there is more than one is that if the student forgets it in the test, he has no alternative. But students always prefer to learn only one method, if it is efficient, rather than learning two and risking confusion. By the law of least effort, they say that if one method already solves the situation, the other is an expendable thing.

Nines. One method that retired in the introduction of Modern Mathematics, by the good side, which was the just demand of rigor, was the infamous "proof of nines", that was the operation of "casting out nines" from the results. Few teachers missed it.

The arithmetic feature used to construct the "casting out nines" was not lost because it continues to be used in the criterion of divisibility by 3, 6 and 9, and consists of adding the numerals that make up the number (if the result is 3 or multiple of 3, the number is divisible by 3, if, besides this, the addition is even, it is divisible by 6, and if, regardless of odd or even, the result is 9 or multiple of 9, the number is divisible by 9).

It worked like this: in the case of addition, the sum of the numbers of the result was summed up, taking nine out, that is, discounting 9 each time the sum exceeded or reached that number. If the sum was 15, we discounted 9 and got to 6. Then the apprentice repeated the operation with the numbers of the two plots, summing all the numbers at once, as if the two were a single number. If the "out nines" of the result gave 6 and that of the parcels also gave 6, one recklessly believed that the account was correct.

If the "nines out" of the parcels was other value, 8, for example, then surely the account was wrong (or was wrong the account made at the time of taking the proof). In case the value is coincident, nothing done! There was no "proof" that the account was correct.

Where is the fault? What happens is that if the sum of the figures of the result gave 15 and the "nines out" resulted in 6, the sum of the plots could have ended in 24, which also has "nines out" equal to 6 (just carry out 2+4). A result that has given "nines out" equal to 1 may have come from 1, 10, 19, 28, up to 91 (the following from 100 onwards should be repetition). If the sum total in the result gave 82 (8+2 = 10 and 1+0 = 1), the sum in the plots would also have to reach 82, to coincide. As the comparison of the value with "nines out" did not guarantee certainty for the account, the method was condemned to the museum of inconsistencies.

Zero. There is no problem when the teacher says informally "top number" and "bottom number" when referring to numerator and denominator, because there is no risk of confusion. He should just keep in mind that students should get accustomed, little by little, to standardized names, so that they always understand the statements. But there are nicknames that should be avoided at all costs. For example, one should never call "zero" of "nothing". When someone says "nothing", he is dealing much more with the idea of "emptiness", the condition of the set that has no elements, than the idea of zero. Zero is not the void, but the cardinality of the void, that is, the number of elements contained in the set without elements.

Instead of considering the number 0 as the center of the sequence of relative integers (integers with negative and positive values), we could agree that the center is any other number, only complicating the accounts a little more. This was what happened in the geocentric system, which considered the Earth as the center of the solar system. The accounts simply got trickier. Thus, 0 is nothing special in relation to the other numbers because it is called zero, but because it has been agreed as the center of symmetry of the whole numbers, below which we only have negative numbers and above which we only have positive values. Since 0 was chosen to play this role, it became a very special number.

In addition to serving to mark the position in the houses without positive figures in the positional system, it is the only number that can receive at the same time positive sign and negative sign, without it being positive or negative, so that -0 and +0 are the same number, 0. Later the student learns that 0 is the only function that is both even and odd, precisely because it can receive a positive signal and a negative signal without changing its value. This fact gives 0 an advantage over the other numbers. But there is a problem that causes this number to carry a disability.

It is about the motive that leads students to develop a certain fear of this special number. As stated above, one cannot divide a positive number by 0, not even a negative number. Whoever tries to divide by 0 a number other than 0, will never get it. Because? When we multiply 2 by 3, we get 6. If we want to know by whom the number 2 has been multiplied to give a result 6, it is enough to do the inverse operation by dividing 6 by 2, obtaining 3. Thus, when we multiply 2 by a number and obtain a certain result, we just divide this result by 2 and find out what this factor was. If the result obtained is different from 0, we will never find a factor equal to 0. If the factor is 0, the result can only be 0. And dividing 0 by 2 we get the other factor, which is 0. Suppose we did not know the factor 2, but we knew the result 0. To find out, we try to divide the result 0 by the other factor, which we already know to be 0. Dividing 0 by 0 we did not find the value 2. This is because any number multiplied by 0 gives this 0 of the result. In place of 2 it could be 3, 4, 300, or even some negative value. Since 0 divided by 0 can give any value, we say that the division of 0 by 0 is an indeterminacy.

If a student multiplied 2 by 0 and got a number other than 0, a 6, for example, the teacher will ask him to do the verification, by reverse operation. He will divide 6 by 2 and he will see that the factor is 3, not this 0 that he put there. And to get factor 2, it is not by 0 that he divides 6, but by 3. Dividing 6 by 0 he will never get anywhere. He should never divide any number other than 0 by divisor 0. The number 0 is no divisor of any number. In contrast, the number 1 is a divisor of any integer, as we have seen. But 0 also has an advantage on this ground: As we have seen now, it is a multiple of any other number. The value 0 divided by any value other than it has a clear and immediate result: 0.

The number 0, therefore, has everything good, except this deficiency of not being able to be divisor of another number. No other number can be divided by 0. Another problem is that if the employee worked the entire week, he will not accept to receive 0 as salary, and if the student has correctly answered questions of the test, he will not accept to receive grade 0. But in this case there is no fault of this special number.

Signals. In the fourth century BC, Plato conceived the idea of negative numbers, but only after a millennium, with the mathematician and astronomer Brahmagupta, in seventh-century, in India, the use of such numbers finally become functional.

It is necessary that the learner first acquire mastery of the absolute numbers, that is, of the numbers with no negative sign, so that he can take advantage of the use of relative numbers. That is why we do not speak of a negative number in the first years of elementary school. At the time of the sixth grade, or of the seventh, with pre-adolescents of 11 or 12 years-old, it is that the subject happens to be studied.

Learning to use signals does not cause difficulties, but the student needs to exercise a lot to avoid making mistakes, either through forgetfulness or lack of attention. One curiosity is that, since boys are less attentive at this stage than girls, they take eight times as much time as the girls do to acquire the necessary training on the subject. Boys, who are more eager to finish their homework and go play, should therefore keep in mind that they need more training. Boys and girls, when they reach the required domain in dealing with the signs, are equally apt in the Arithmetic of relative numbers and in the initial Algebra.

Andrew is already well prepared for this, but he spent more time on the subject than his classmates Adriana and Julia. He learned that when two negative numbers are added, the result is a negative number. Thus, adding -30 to -20 means reaching the number -50. This sum is written as (-30)+(-20). He has previously learned that +(-20) and -(+20) mean the same thing and are worth -20. Hence, to subtract (-40)-(+20) he changes this expression to (-40)+(-20), obtaining -60. This maneuver of transforming subtraction into addition is called algebraic sum. Everything happens as if there was only addition, which is done with both negative and positive numbers. The addition operation done with numbers of exchanged signals, however, obeys the idea of subtraction. So, adding +10 to -50, that is, doing (+10)+(-50), results in -40. In practical terms, if someone has won 10 points in the game and then lost 50, his balance is a loss of 40 points. If he had won 60 points and lost 40 then his balance would be +20 as a result of the operation (+60)+(-20).

If we are at minus 6 degrees and the radio says it has dropped another 3 degrees, what we have is (-6) plus (-3), or (-6)+(-3), which results in -9 , or negative 9 degrees.

Since -(+1) and +(-1) mean -1, Andrew has already learned that he can do the algebraic sum without using parentheses, eliminating them when they appear explicitly. He also knows that +(+1) means +1, but what will be -(-1)? The teacher had explained that when we apply the opposite (negative) operator, the minus sign, at a given value, the representation of the value on the numerical scale, or on the number line, jumps to the symmetric point. When we draw a horizontal number line in the notebook, a ruler graduated with 0 in the center, having negatives before and positives after, the number 1 occupies the next integer position to the right of 0, and the number -1 occupies the first position to the left, that is, the position symmetrical to that of the number 1, which is the same as +1. If we apply the opposite operator to this +1, doing -(+1), we jump to its symmetric, -1. If we apply the operator to -1 by doing -(-1), we jump to the opposite side, +1. This fact works for any integer. Hence, -(-8), the opposite of -8, gives +8. Knowing these things, Andrew saw how easy it was to eliminate parentheses in algebraic sum. An operation like (+10)-(+5)+(-20) can be written as 10-5-20, without going through the (+10)+(-5)+(-20) phase. At first he made the operation following the appearance of the numbers, keeping an eye on the resulting signal at each step, but then realized that it is more practical to add up the positive values, then add the negative values. The result is the subtraction count between the two totals. In the case of 10-5-20 we have +10 and -25, whose algebraic sum, in this case a subtraction, gives -15. If it has something like 4-7-2+8-5-3+9, it gets +21-17, which results in +4.

In the passage to the learning of the signs of multiplication the teacher had explained that the symbol of parenthesis has meaning of "times". Just as the preposition "of" between fractions means "times", parenthesis between numbers also represents this operation. The count 4*(-3), four times -3, can be written as only 4(-3). The result is obviously -12, because it is the -3 number added 4 times. Since +4(-3) is giving -12, there is an example showing that "plus times minus" gives "minus". "Minus times plus" also gives "minus", because it is just a matter of applying commutative property. Multiplying two numbers of exchanged signals always gives the "minus" signal. Multiplying two "plus" signal numbers will certainly result in "plus". The negative cannot fall from the sky. The professor, however, had shown something that was fixed in Andrew's memory. The number 1 is a neutral element of multiplication, so multiplying 1 by a number is the same as not multiplying: +1*8 is the same as +8*1, which gives itself 8, or +8. If we have -1*8, this gives -8*1, which is -8. At -1* (-1) we can write -(-1) *1, or -(-1), which is +1, symmetric of -1. Thus, for -1*(-8), we do -(-8)*1, which gives +8. The conclusion, by numerical examples, is that "minus times minus" gives "plus". The teacher had written on the blackboard the block that follows below.

> For *multiplying any signs,*
> *Plus times minus is minus,*
> *Minus times minus, plus,*
> *Now the things get clear and fine.*

In doing division operation, the teacher explained that the rule of signs is the same as multiplication, since division is the multiplication by the inverse of the number.

Andrew no longer forgot this information. He just needed to practice.

There is, however, a problem that many students face in this topic, and it refers to the language that teachers use. Most teachers try to teach the multiplication sign scheme by replacing the word "times" with the word "with". By saying "minus with minus", the teacher refers the student to the addition account. And the learner will have great difficulty in distinguishing the signs of addition and multiplication.

So that there is no confusion, the teacher should say always "minus times minus", without ever using the "with" preposition when trying to multiply. And the student himself must train his mind with the phrase "minus times minus", fleeing from this "with", even if the teacher remains rebellious, uttering that.

Expressions. When the commitment to primary education began to fade, many teachers gradually began to ignore an essential topic in the learning of Arithmetic, which is that of numerical expressions. Before "Modern" Mathematics the subject was called "arithmetical expressions". Then it received the shorter, and apparently less scary, name of "numerical expressions". The topic is important for two main reasons: it works as a closure of the Arithmetic theme, joining all the pieces in a single process, and serves as training for Algebra, which the student has to learn shortly after. This training is due to the new learning of the use of "clasps", which are the parentheses, brackets and braces used to link one or several operations in the same exercise.

Before we speak of larger expressions, let us speak of the simplest ones, which are those that contain only operations between numbers. For example, 8-5+4, which has result 7. What if we mix addition and multiplication? Then we have to obey the order of operations, which comes in the opposite sequence to that of the chronology of learning: first we solve multiplications (and their inverses, divisions), and then we pass on to additions, with their inverses, subtractions.

An expression like -5+3*8-2*3+4:2 cannot start with the -5 count, because then we see that it is being added to a product. We have to repeat this -5 and do the multiplication and division operations. We will have -5+24-6+2. Now it is easy, just we add the positives and then add the negatives: 26-11. This gives 15.

When we write above (+10)-(+5)+(-20), we show a numeric expression, in one of the very simple forms, with only parentheses, and inside each only a number. We can enlarge the idea using parentheses, but adding operations inside them. For example, (-4+11) + (-7) - (+5-8). There is more than one way, and, without using the rule of numerical expressions, there is a temptation to eliminate the parentheses at once. It is possible, but it is not the recommended method here, because the moment is to train the use of "clasps". The resolution technique requires that, before we delete the parentheses, we do the operations inside them (I remind you that this way of doing is not mandatory, but recommended in the technique we are presenting). In the example, -4+11 will result in +7 (just subtract from back to front). The whole expression will be: (+7)+(-7)-(-3). Now that we have done the operations inside the pairs of parentheses, we can dispense them, with 7-7+3 (the "+" sign in the first 7 could have continued, but it is totally dispensable at the beginning of the expression). By Andrew's preferred method, we get 10-7, which gives 3. We could also, before adding the positives and adding the negatives, eliminate, by cancellation, the symmetric numbers. Cutting 7 with -7 would have the result 3.

The initial example with only parentheses did not come out randomly. The priority is with them. When we have in the expression parentheses, brackets and braces, first we have to solve the contents of the parentheses, (...), beginning with the operations inside them. Then, the priority goes to the brackets, [...], and finally to the braces, {...}.

If we have 18-3[4 + 5(6-8)] + 2, before thinking about the brackets we have to just repeat them, because the priority is to deal with parentheses. We get 18-3[4 + 5(-2)] + 2. Now we do the operations inside the brackets, because this pair of parentheses is just to save the signal of the internal number, -2. We will have 18-3[4-10] + 2, then 18-3[-6] + 2, which gives 18+18 +2 = 38.

Finally, let us look at an example with braces. Let us solve the expression {3-[5+(8-3)]} - 1. First, parentheses: {3 - [5+5]} - 1. This gives {3-10} - 1. Solving the braces, we have -7-1 = -8.

While knowing the whole technique for expressions with integers, the students will apply it without complications when it has expressions of power and roots, first, and with fractions, later.

Those who had resistance to this topic put the nickname "wagon" to any numerical expression exercise. This nickname has smell of arithmophobia, for sure.

7. Peano's axioms

The rigor in mathematical language came from the Set Theory, but not just from it. It was a requirement of time. Professors David Hilbert, from Germany, and Giuseppe Peano, from Italy, both very active in the early twentieth century, were largely responsible for this change, whether or not they used set notation.

Jules-Henri Poincaré wrote that the step that in Mathematics corresponds to the Darwinian Theory of Evolution, of Biology, is the axiom of Finite Induction, by Giuseppe Peano. It is simply the last of the five axioms of natural numbers, which Peano presented to the world in 1889, in a thirty-page pamphlet called "New method of exposing the principles of Arithmetic" (*Arithmetices principia, nova methodo exposita*).

Peano was a professor of Mathematics, but he was also concerned with linguistic issues. In 1903 he launched the proposal of a modern Latin, "Latino sine flexione", as an alternative to the Esperanto language. Peano's simplified Latin was called Interlingua by Peano. After many changes and ramifications, made by other scholars, the version of linguist Alexander Gode, the Interlingua-IALA, of 1951, which today is practiced by groups of scholars in almost all the countries of the world, was consolidated as the most promising (IALA is the acronym of International Auxiliary Language Association). Example of a sentence in Interlingua-IALA: "Nos necessita studiar plus Mathematica ora" (We need to study more Mathematics now).

In Mathematics, Peano's five axioms, among many other contributions of him, are as follows:

1) Zero is a natural number.

2) If **n** is a natural number, then the successor of **n** is also a natural number.

3) Zero is no successor to any natural number.

4) If two natural numbers **m** and **n** have the same successor, then **m** and **n** are the same number.

5) If a first number belongs to a set and, given any natural number, the successor of that number also belongs to the set, then all natural numbers belong to that set.

At that time there was not the trend among mathematicians to consider 0 as a natural number, so that Peano's text spoke in 1 as the smallest of the natural numbers, not 0.

Alternatively to Peano's axiomatic approach, Bertrand Russell developed a presentation of natural numbers through definitions, based on biunivocal correspondences between sets, which is an idea he located in an article by Gottlob Frege. The number 3, for example, is defined as the common property among all sets that have that amount of biunivocal correspondences, which we call 3. This, in fact, looks like a tautology, a definition that revolves around itself.

Peano had no distrust of Set Theory, like Poincaré had, so much so that he was the one who created the symbols of "it belongs to", "it contains" and "it is contained in", embodied in basic Mathematics. The "Peano's five axioms", unlike Frege's approach, form a clearer and simpler presentation for the idea of number.

Induction. More than that, they bring the strength of the Principle of Finite Induction, also called Postulate of Mathematical Induction.

A set of numbers is defined by a property. If it is randomly formed, this is its nature. Otherwise, some other criterion exists. Thus, when the fifth axiom speaks on set, it takes into account the property that defines it. So to say that all natural numbers belong to this set is to ensure that the property holds for all natural numbers.

The understanding of the meaning of the axiom is not immediate. Many authors imagine that it is necessary to ensure that the property is valid from a first number up to such a given number. It is an unnecessary addition to Peano's statement. What happens is that if we fix the number 1 as that first number (we must start by proving that the property is valid to this value), and take a number **k** to prove that the property holds for the successor k+1, we must take into account that if this **k** is 236, the validity was proved to 237. If it is 900, it has been proved to 901. Now, with much more reason it is guaranteed that of 1 it is worth for 2, of 2 it is worth for 3, and so onwards. The demonstration, without needing to embed any inequality, guarantees that the property holds for any natural number, just as formulated by Peano.

Many have had difficulty absorbing the concept also because of the mechanical induction, which is recognized as an inconsistent reasoning. If a person goes to a beach house and there it rains every day at 4 pm during the first 15 days of the stay, there is no guarantee that the next day it will rain. If the succession we are dealing with refers to the passing of days, or hours, one cannot prove anything by induction. In Peano's fifth axiom there is no reference to the passage of time. The sequence is of the natural numbers, simply. When the property is valid from 1 to 2, in the same instant it is worth one million to one million and one. There is no passage of time. Those who imagined this passage should give up that belief.

Demonstrations. I will not show the demonstration here, but I give you an example of how one can use the axiom to do a proof. It is known that the series of odd numbers, from the number 1, always results in the square of the value of the position. If we add three numbers, the total is 9. If we add four numbers, we will have 16. Let us see: $1 + 3 + 5$ results in 9, while $1 + 3 + 5 + 7 + 9$ (five numbers) gives 25. The number 1, which only has one position, is worth the square of 1. If we take **k** numbers, we are assuming that the sum will give k^2. Our proof will end when we prove that by taking the next position, of k+1 numbers, the sum will give the square of that k+1. The work is simple and I for myself have done it many times in classroom.

Why should anyone indulge in this work of demonstrating property? Because this is the work of the mathematician. Before Pascal invented the calculating machine, mathematicians made their living doing accounts, drawing astral maps and making demonstrations. Now, with the calculator, the four operations are learned in school, but are performed in the various professions with the use of machines. The astral map, since the French Academy of Sciences removed Astrology from the list of scrutinizing sciences, that is, from the sciences themselves, mathematicians were far from it. The only job left was to make demonstrations.

Why is it important to demonstrate properties? Because this is science. To know if a future 200-story building will stand erect after it is built, we did not build it first and then check the hypothesis. We do the necessary calculations and demonstrations first. To send a manned spacecraft to the Moon, we do not launch the vehicle up to see if it will reach the natural satellite, but we first do all the calculations and demonstrations necessary for the planned trip.

Without the practice of demonstrations, we would make canoes, hooks, fair carts, chairs, shelves, clothes, hovels and everything that is within reach of our hands. Things that transcend this, such as transatlantic, trains, paved roads, hydroelectric, we could even do, but always taking colossal losses, by mistake of planning. The failures would be so great that only the insane would continue to insist on these undertakings.

Small objects that our fingers cannot assemble and dismantle alone, such as small clocks, cell phones, cell radios, light bulbs, none of this would be built without previous demonstrations.

Education systems need to be taken seriously again so that junior high school students can learn to do classroom demonstrations once more. Without it all science loses a lot, starting by wasting the brains of youth.

Arithmophobia: How to Heal the Horror of Mathematics

If teens finish their third year of High School without ever having demonstrated any property, any theorem, all of the math they have learned is harmed. It is something that has a little kinship with Mathematics, but it is not Mathematics. The intellectual foundation of the Academy of Plato is based more on the demonstrations than he did that in his texts of Philosophy. Creativity in making demonstrations is as important as creativity in enunciating new propositions. Many mathematicians have gained renown only by demonstrating the validity or falsity of facts that others have enunciated.

The national High School exam in Brazil, "Enem", has never required a demonstration, once the test is multiple choice. This means that there was never actually Mathematics in that contest. For classification purposes it works, but it has given youth and school system a misconception of the meaning of science.

Demonstration is seven-headed bug? Not even a little. It requires trekking, certainly. A student of Greek is not required to keep a conversation in that language after two months of classes in the subject. For the same reason, a nine-year-old student is not expected to be able to demonstrate geometric facts. It is not impossible, but it is impertinent. As much as a 10-year-old boy studies aviation, passengers in a company are not expected to accept travel on a large jet flown by him. The child will do demonstrations from the age of 13, based on exercises, if the education system becomes serious again. And at age of 17 the student will be ready to demonstrate unpublished facts.

Euclid. Euclid of Alexandria, third century BC, professor of the Museum of Alexandria, the Greek equivalent to the Athens Academy, compiled all the Geometry of his time in his compendium "Elements", of 13 books. When he died, in 265 BC, the future librarian Eratosthenes was 10 years old and still lived in another city, Cyrene. Inside his Geometry was all the well-known Arithmetic, as well as all Algebra, which was not yet symbolic, but discursive.

It was remembered now because the assertive known as Euclid's Theorem is easy to understand and can be used here as an example of demonstration. The terms used here are not those of Euclid nor are those used in textbooks. The idea is what matters. The proposition, also called the Prime Infinitude Theorem, states that there is no prime number which is the largest of all. The proof is made by absurdity, that is, by contradiction. In this model of proof, based on Aristotle's binary Logic, we assume that what counts is the opposite of the proposition, and then prove that our assumption is absurd, concluding that the valid statement is the original.

Euclid supposed, absurdly, that there is a "top" prime number, greater than all other primes, which are therefore in finite quantity. There is a product of all these prime numbers, a number K, which can be decomposed into all those prime factors. Then he constructed a successor to that compound number, K+1, which we can call **q**. We now have two possible situations for this number. (A) It is a prime number, and this disassembles everything we assumed before, since it is the successor of that entire product, greater, therefore, than the greatest prime number we had imagined before. (B) It is composed, which means that, as the existing prime numbers are inferior to it, one of these prime numbers is divisor of it, that is, it is in its decomposition into prime factors. Now, **q**, or K+1, has a factor that is some prime **p**. Since **q** and K+1 are equal, q = K+1, we immediately see that q-K = 1. The prime **p** is a factor of **q** and also of K, which means that it is a divisor of the difference q-K (if the reader is accustomed to algebraic factorization, **p** is taken in evidence). Great! Great? None of this! q-K is 1 and the number 1 has no prime divisors. Contradiction!

What Euclid has shown is that if there were a greater fixed prime number, the product K of it and all the previous primes would have a successor **q** that would be neither prime nor compound, i. e., **q** cannot exist. Therefore, the greatest prime number cannot exist as well.

8. Strong motives

I try to present to you now, reader, strong practical, historical, philosophical, aesthetic, and other reasons to make the student enjoy Mathematics.

Thales of Miletus, seventh century BC, was the great inventor of the laboratory science, when he made the first demonstrations in Geometry. The great researchers of Greece, from that time, began to devote a certain contempt for Arithmetic, understood as an activity of merchants, not of philosophers (before Galileo, sixteenth century, who made science was a philosopher, and from there until the nineteenth century, a scientist was also a philosopher at the same time; in the emergence of Pure Mathematics with George Boole, in 1848, the separation of philosophers and scientists began, but it is a separation that should never have taken place; this is what I think).

Prejudices. Currently, there are still some great mathematicians who twist their noses in front of Arithmetic, Geometry and Logic researchers, who make up the fundamentals area. But they are people who cultivate prejudices and are not good examples for citizenship.

The first book of the great Condorcet (1743-1794) I took to read in the college library made me strange at first. It was a book of few pages, in publication of more than a century, dealing with learning of basic operations of integers. He taught patiently techniques for the young student to learn to make his first accounts. I immediately thought of the fact that great doctors I knew would not bother writing a handbook for rudimentary students. Soon the scare passed, when I reasoned: I was before Condorcet, the mathematical who as a parliamentarian was responsible for the authorship of the law that implanted the French public school. There is a natural tendency, which will contaminate the mathematical researcher, if he is not careful, to value some areas and consider others as a minor thing. If Gauss had such a reserve, he would not have been the greatest mathematician of his time.

Prejudice against basic Arithmetic not only contributes to hinder the healing of arithmophobia victims but also shows a nephelibata behavior (of those who live in the clouds), once Arithmetic is the oldest and deepest foundation of all Mathematics, prior even to Geometry and Logic. The attitude of contempt is equivalent to that of a citizen who boasts that his school has not helped him in his learning path.

Until the beginning of the 21st century, Brazilian students studied English as a subject from the sixth grade of the fundamental course to the last year of the High School, for a total of seven years in cohabitation with the matter. Today they start earlier, in large part of the municipalities. Already I once saw on TV the interview of a Brazilian who emigrated to the United States in which he said that he left Brazil with his full secondary education, but not knowing a single word of the English language, and that in two months in the new country he was speaking fluently. What this liar and biased does not know is that an adult does not learn a new language in just two months if he does not have a basis. When he confessed, bragging, he learned in two months, he disproved the earlier statement, that he had not learned any word of the language. Certainly he did not arrive in that country showing fluency in the language, but he had the necessary baggage to take, in a few weeks, the step of speech independence.

This silly prejudice against the previous basis of learning also occurs in relation to Mathematics. As there are several weekly classes, over the years, the student accumulates knowledge, even when he is not clear about what he is doing. At any given time, in the sixth grade, the seventh, or in some later year, the concepts begin to make sense.

A colleague of mine who had already completed an undergraduate degree in Mathematics in an ordinary college, joined the undergraduate program again at USP (University of Sao Paulo). He told me that, only when he started the second year of USP, he begin to realize what Mathematics is. This means that until then he accumulated concepts, obtained marks in tests, but without being aware of what he was doing.

A teacher of mine, in this course at USP, said in class one day that only at the age of 17 she become aware of her own existence, that is, of the individuation process, in Jung's conceptualization. She could be referring to the consciousness of learning in Mathematics, since this would be her future career.

If the student attends a serious school, committed to content development and evaluation, between 12 and 13 years old he will already have clearness about the meaning of the mathematical science. But, as we have seen in the case of my colleague, he can finish graduation in the area with no idea of what it represents.

Occam. Is it really so hard? Everything looks like as relative. If the student is studious, attentive, and if his school is efficient, facility comes as a consequence. If someone thinks he can burn steps, or can learn without effort, math will never be easy.

Arithmophobia: How to Heal the Horror of Mathematics

What Mathematics will always be is a simple thing. About being easy, not always.

According to the Principle of Parsimony, famous Occam's Razor, Mathematics has always been built on the simpler path. Between two methods equally rich in ideas and equally efficient to solve the same question, if one is more complicated than the other, the complicated one falls into abandonment.

In addition, all science always seeks simplicity. Occam's Razor is the principle that recommends that between two explanations for a fact, under equivalent conditions, the one that has more chance of being true is the simpler one. Regardless of this, when research is carried out, it seeks to discover new facts, but also to simplify the explanations and understandings that already exist.

The student should not be afraid to move forward in classroom content because, while knowledge accumulates, everything tends to be clearer and simpler as more basic topics are left behind.

For example, Mathematics teachers know that fractional addition learning, this rite of passage without which all middle and junior learning will be pure deception in the student's head, occurs most easily at the end of the primary course or, early in the beginning of the third triennium, the seventh grade. Obviously, until the age of 16 or 17 the student still has a lot of openness to learning the subject, albeit with a smaller gain, as he wasted a precious phase, in junior high school and early middle school, to exercise on the subject.

But unfortunate of the student who enters a demanding course of Engineering, Physics, Meteorology or Economics without having learned fractions before. It is not impossible to learn, but the suffering will be great if he does try.

Another colleague of mine taught Differential and Integral Calculus in a median college, later transformed into a university, in the city of Sao Paulo. A very studious student had studied the subject with him five times. He surpassed all other matters and was going ahead, but not in Calculus, since she was not certified in Calculus I. As my colleague was rigid in evaluation, there was no fraud or charity. One had to learn and pass, or missed the questions in the test and continued without promotion. In this fifth time the girl studied the subject with him, he decided to investigate in depth what was her problem, once she learned the concepts well, passed in other subjects, but with him she did not advance. And she already taught math in basic schools. He discovered, and then checked face-to-face, that she had no difficulty in almost anything. The problem was in fractions. He simply did not imagine that a college girl might be locked in her course because she did not master fractional operations in the fifth grade - if the school unit offers the subject for first time only in the sixth or seventh grade, it is late.

Since she was already a Mathematics teacher, she taught the subject without knowing the most crucial subject of Arithmetic at the fundamental level. Surely she skipped the pages of the book. And she probably figured that the topic was not important. Likewise, many other teachers with gaps in their own learning are working in the education networks. That girl faced my colleague's sieve. How many in the world were not approved without knowledge? Students are at constant risk of poor teaching if the evaluation is only in the hand of the teacher of the occasion. A barrier to learning because of a knowledge gap may be the trigger for a crisis, often silent, to start arithmophobia.

Pascal. I return to Blaise Pascal (1623-1662) because there is an episode of youth that is said to have defined his future as a researcher. At age 17, one day he had a very sharp toothache, and he did not know what else to do. Extraction of teeth at that time was without anesthesia and was something that was done in the last case.

In desperation, the young Pascal took a chore to do. They were math exercises. Even though he felt pain, he, as a well-disciplined boy, did not escape responsibilities. His teacher, in the various subjects of knowledge, was his own father, in house. His mother had died in 1626, when he was only three years old. (Here there is a common point between Pascal, inventor of the calculator, and Ada Byron, Countess of Lovelace, who, two centuries later invented computer programming: Both lived the shortage of one of the parents, for Lord Byron, for leading a bohemian and dissolute life, was forbidden by his mother-in-law to see his wife and daughter as soon as she was born, so that Ada did not know her father personally, only having the consolation of having, while having died of cancer at 36 years old, her bones resting beside the paternal ones, since she demanded to be buried next to the grave of Byron.)

He continued to do the exercises, and in a few minutes he noticed that the toothache disappeared. So he continued his workout and went further, doing much more than what he needed. If he stopped, the pain might come back. He did not come back, at least that day.

Pascal's conclusion was that Mathematics has a medicinal, psychosomatic power (eventually this word did not yet exist). And he decided that he would be a lifelong mathematician.

Arithmophobia: How to Heal the Horror of Mathematics

Retrieving experiments from Evangelist Torricelli, the most famous student of Galileo, in 1646 Pascal demonstrated the existence of the vacuum. In 1648 he enunciated what later became known as Pascal's Principle, "All pressure exerted on a fluid (any liquid or gas) spreads throughout the substance uniformly". At the same time he established the law of communicating vessels.

During a trip in 1652 with some nobles, among whom Knight of Méré (Antoine Gombaud), he was influenced by this one and happened to dedicate himself also to philosophical subjects. But he did not diminish his choice for the development of Mathematics, which, at that time, included Physics. In 1653 he wrote a complete treatise on Hydrostatics. That same year, instigated by discussions with Knight of Méré, he dedicated himself to the study of Probabilities and developed the first works in Game Theory, sharing ideas with Pierre de Fermat.

In 1654 he published the Treaty of the Arithmetic Triangle, presenting what happened to be called the Pascal Triangle, in which each line is formed by all combinations (binomial ones) of the line position minus unity. Thus, in the first row there is only one combination, with n=0, since the line has position 1 and the unity is discounted. In the second row, with n=1, we have the combinations with 0 and 1. Then, with n=2, we have the combinations with 0, with 1 and with 2, and so on. With the accounts solved, the successive lines are 1, 1 1, 1 2 1, 1 3 3 1, following up the line number one wants. To construct new lines without having to compute the combinations, simply we repeat the number 1, which is the beginning of each line, and add the values from the previous line two to two. Thus, to construct the fifth line from the fourth, 1 3 3 1, simply write 1 and add 1+3, then 3+3, then 3+1, then 1+0 (the 0 is not written, but we imagine it). The line will be 1 4 6 4 1. As for the sixth line, reader, can you build it? (Combination of **n** with **p** elements is the quantity of sets with **p** elements taken from the total set; for example, in the set {a, b, c}, of cardinality n=3, if p=2, we have 3 combinations, which are {a, b}, {a, c} and {b, c}.)

In his work on the "arithmetic triangle" Pascal happened to present the Principle of Mathematical Induction, embedded in his triangle, as a method of demonstration, but the mathematical world remained unaware of the power of that instrument until it was made explicit in the Peano's five axioms.

After dedicating himself to philosophical and theological writings, like the book *Thoughts*, he resumed mathematical research, launching in 1659 the *Treatise on the Sins of Circle Quadrants*. Leibniz later recognized that his inspiration for the creation of Differential Calculus was in this Pascal text.

In addition to the calculator, Pascal was the founder of collective transportation, formulating a large type of carriage, in which Parisians paid 50 cents to travel around the city. From what is portrayed in Rosselini's film "Pascal", he was sick, on the deathbed, when they came to tell him that the novelty, his carriage, had been implemented.

Wonders. I have never seen anyone who thinks Lewis Carroll's book, *Alice in Wonderland*, is a minor work. Unlike many other works of the genre, which are compilations or redesigns of traditional popular stories, Lewis Carroll's best-known book is the product of his imagination. As it is said in the present academic circles, it is authorial work.

Just as Pascal, a great mathematician of the seventeenth century, made a great contribution to the public service, which was the vehicle of collective transport, Charles Lutwidge Dodgson (1833-1898), an English mathematician, left as an important legacy to the literature of entertainment his book Alice, of 1865, for Lewis Carroll is simply the literary pseudonym of that mathematician. Trying to Latinize his name, he came to Ludovicus Carolus. As this did not result in anything too loud, he found an English version for this name: Lewis Carroll. After publishing poems and short stories with his original name, he published the poem *Solitude* under the new pseudonym, in 1856, and saw that it was a good idea to keep the practice, perhaps to dissociate this work from his activity as professor of Mathematics at the University of Oxford, and also from his religious work, once he was as Anglican deacon.

With the success of Alice, he published shortly afterwards *Through the Looking Glass and what Alice Found There*, as a continuation. In 1876 he published *The Game of Logic*, and in 1879 he launched *Euclid and His Modern Rivals*, a discussion of the didactic of Geometry, in the form of a play. His last literary work was *Sylvia and Bruno*, in two parts, published in 1889 and 1893. He had fun inventing games and discussing paradoxes. Many say he is the inventor of the barber's paradox, used by Bertrand Russell to question the then-intended scope of Set Theory: In a given realm the barbers were given the order to shave all the people, who should not shave themselves; in one district there was only one barber, who complained that he could not be shaved, once he was the only barber there and no one could shave himself. Thus, not all could be shaved, contrary to what the order of the monarch intended.

From Charles Dodgson is also the creation of the double entry table (Carroll diagram). Some crazy people swear that this was on a multiplication table invented by Pythagoras, which crosses line number with column number and shows the result of the product. Before the Internet these false authorships were already very common.

As a notable contribution to the study of numbers, he published in 1867 the study *An Elementary Theory of Determinants*, in which he presented the conditions for a system of linear equations to have nontrivial solutions, that is, solutions with nonzero values. A linear equation, if you do not remember, is a first-degree equation of one (x) or more (x, y, z,...) variables, which cannot be raised to exponents other than the number 1 He also invented cryptography methods and made contributions to electoral calculations.

Lewis Carroll was also a photographer and his preference in this craft was to portray girls, almost always daughters of his friends and neighbors. Among them was Alice Liddell, which led to the suspicion that his book Alice had based on actual history with her, a fact he denied. He just drew on her name, which no one can deny is a beautiful name. This preference for portraying girls caused him to face insinuations of having a pedophile bias.

People who create important novelties and solve problems should always be warned against slander, for the mountebanks and mythomaniacs are always on the lookout for a career in the work of others, or, more seriously, on the reputation of others. Socrates and Giordano Bruno are not isolated or casual victims. So it was that Lewis Carroll has had to go through charges of being the identity of Jack the Ripper, the anonymous infamous serial killer from London in 1888. The person who made the allegations claimed that metaphorical phrases of works previously published by the author of Alice represented clues and passwords for the crimes he would commit under the Jack's dark and mysterious figure. To this day, nobody knows for sure who Jack was, but bringing charges against Lewis Carroll based on alleged keys left in his books is a clear manifestation of some delusional disorder (according to the book *Naming Jack The Ripper*, by Russell Edwards, recent studies, made 120 years after Jack's female deaths, through DNA tests in the shawl of one of the victims, and in the sample of semen found there, pointed out Jack's identity as a Russian immigrant named Aaron Kosminski, an unemployed hairdresser who was 25 at the time).

Proportion. Plato wrote that the first step in doing science is classification. The precept is undeniable, but everything would stop at that point if we did not know other steps. When a farmer separates his products into lots to sell at the fair, he is just sorting. What follows in his activity does not have much relation to doing science. The science began with Thales of Miletus because, after classifying facts, he used proportion resources. At the time he was not aware that he was creating a new method of building knowledge, which was later called the scientific method, so he has the title of Father of Philosophy, not Father of Science.

The in-depth Theory of Proportions was developed by Eudoxus of Cnidus three centuries later (he lived between 390 BC and 337 BC), but what Thales knew was sufficient for his purpose in his time.

The proposition known as the Thales Theorem ensures that if we draw a line transverse to a bundle of parallel lines, in any other line transverse to that bundle the resulting segments will be proportional to the corresponding segments of the first transverse.

The simplest case is that of three parallel lines, preferably with different distances between one and the other. A first transverse intercepts the three parallels producing between them segments of sizes **a** and **b**, from top to bottom on the paper. Then we draw, in another position and with different angle, a second transversal to the same parallels. Since the angle is distinct, the sizes of the produced segments are different, and let us call them **c** and **d**, also from top to bottom. What the Thales Theorem guarantees is that the two ratios a/b and c/d are equal, that is, the proportion a/b=c/d is valid.

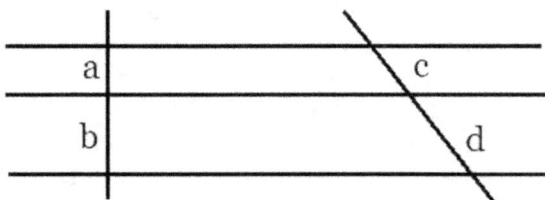

Thales: Bundle of parallels and two transversal lines

To find one of the values of the proportion, when incognito, a fourth proportional (the letter **d**), a third proportional (the letter **c**), or any of the four values having the other three, we usually apply the technique that came from the Arabs of the Middle Ages, called the "rule of three".

How to find the fourth proportional **x** in the expression 3/4=9/x? Simply we multiply the ends (3 and **x**) and equal them to the product of the means (4 and 9). It is the fundamental property of proportions: *the product of the means is equal to the product of the extremes*. In practice, we always first choose the arm that has the unknown **x**. In this example, 3*x=4*9. To solve it, we do 3x=36. Then it is only divide both members by 3, which means that in the first member will only be 1x, or, simply, **x**, with the value 3 turning a divisor of the second member: x=36/3 , or x = 12.

Now a practical problem. Gabriel bought 4 kilos of lentils for $ 20. At the same price ratio, how many kilos would he have bought with $ 75? Comparing the ratios kilos per kilos and price per price, we will have the assembly: 4/x=20/75. It is to find the second proportional. Beginning with the product of the means, we have 20x=4*75 (when we multiply a number, a coefficient, by a letter, we dismiss the symbol of "times", following René Descartes' suggestion). We get 20x=300, which gives x=300/20, or x = 15. Gabriel would have bought 15 kilos of the product.

What we have seen above are examples of direct simple rule of three. In the latter case, if we increase the volume of money, we increase the amount of kilos of lentil. But let us look at the following situation. We have a farm plot to plow and we know that if we hire 4 employees, they will spend 5 days to sow. If we hire more employees, will they spend more days, or a fewer number? It is clear that they will spend less days. Increasing a value on the first member implies decreasing the corresponding value on the second member. Here we have a inverse simple rule of three case. How do we solve it? We simply set up the problem as in previous cases, and since we know that one of the ratios has inverse growth, we reverse the second ratio. With the numbers above, from 4 employees to 5 days, if we hire 10 employees, how many working days will it take? We make: 4/10=5/x. Since the rule of three is inverse, we reshape to 4/10=x/5. We will have 10*x=4*5, or 10x=20. This gives x=20/10, or x=2 days. It is very important that the ratios be assembled by adding values of the same nature: 4 employees for 10 employees, on the one hand, 5 days for **x** days, on the other.

In a simple rule of three we will always be comparing two ratios, that is, two pairs of elements given in the form of ratio. In a compound rule of three, we have three or more pairs of elements, three or more ratios, which we must transform into two to solve. I leave you, reader, the task of researching the subject in textbooks or on the Internet, if you need to deepen in it, for some competition or another type of examination.

We owe much to the idea of proportion, because with it science began. Without it, we would still be sitting on wooden logs, because a mere chair that the carpenter would build without the use of science and proportion would fall under the body of the first adult sitting on it.

In our urban life, throughout a day of work we come in contact with dozens of different instruments and machines, all of them propitiated by the scientific development. If we count the virtual machines, the number multiplies by another dozen.

Business. It is difficult to devote to science and concomitantly secure time to make money. One cannot imagine Charles Darwin or Albert Einstein doing business and getting rich. But there are cases where the scientist is already in the business world. In the time of Thales he did not earn his living by dedicating himself to science, and he had the profession of a merchant to live. It is an office in which one learns from experience that one should not aim to earn little. If the trader accepts this, any small market crisis leaves him in the negative. So, to be on the safe side, he should always aim for good profit margins in his transactions.

Thales, according to Aristotle, used the knowledge he had to earn good money at a certain stage of life. He did not marry or raise a family, but a single bachelor merchant also needs to be on his guard, preparing against financial downturns. He predicted, using his calculations, that the weather would cause a major breach in the rice crop in Greece. As only he knew this, he began to buy and stock up everything he could from bags of the product. When scarcity came, the man who could supply the markets was the merchant Thales of Miletus, who charged the price he thought reasonable for a man who at that moment had a monopoly on the supply of a commodity of prime necessity. If he was not yet rich, he became rich in that moment.

At the age of 30 his mother began to urge him to get a fiancee and get married. He claimed it was still early. Years later, his mother insisted again. Very busy with business and with philosophical and scientific investigations always, he convinced his mother that her grandchildren could wait. When she saw that perhaps she was losing the time of having her grandchildren, she returned to the charge.

- Thales, my son, I think now is the time to get a wife.
- Ah, Mother!
- Ah, Mother, what?
- Ah, Mother, it's too late for that now.

When he was 62, he left home one night almost early morning, and went walking, staring for a celestial phenomenon that would occur on that date. The girl who was a servant of the house found strange that night's walk of her master and went behind, watching him in the distance. At some point along the way, always looking up, he fell into a dry, very deep pit. Such was the death of this extraordinary man.

Golden. Every form of proportion has a great beauty, no doubt. But there is a kind of proportion that since the times of classical Greece has enthralled architects, mathematicians, engineers, technicians, musicians and artists in general. This is the *golden proportion*.

Franciscan friar Luca Pacioli (1447-1517), a great Florentine mathematician, wrote the first book devoted entirely to that proportion, and intended to baptize it as "The Divine Proportion". He invited Leonardo Da Vinci to illustrate the work and he said that he preferred the work to have a lay title.

At that time, while being urged by the mathematician to find this title, Da Vinci suggested the adjective "golden", in place of that "divine".

After all, what does this proportion refer to, deserving such sublime names? To introduce it, we usually use segments, whose sizes will be the terms of the ratios involved. We take any segment and in it we mark a point that we will call "golden section", as Da Vinci called it. Did we make a point anywhere? Not! Otherwise it would not be the golden point. Within the infinite positions of the segment, the golden ratio point, which performs the golden section, is the one that divides the given segment into two segments of distinct sizes such that the smaller is for the larger of the two as the larger is for the original segment.

And what is the grace of this section? Before any other discoveries, the ancient Greeks already knew that it produces the most harmonious division, the most pleasing to the eye.

If we take a piece of wood, about 40 centimeters long, and we break the piece in the position of the golden ratio point by folding it 90°, that is, at a right angle, then this larger side and this smaller side can represent the length and the width of a laptop, for example. They can also be the length and width of a copybook.

In architectural constructions, it does not matter whether the largest segment is width, as in the Parthenon, or height, as in the UN headquarters in New York, the ratio between the two dimensions should approach the golden number. Number? Yes, because a ratio always corresponds to a certain number, which we will know when we discover the "fraction" equivalent to it with unitary denominator.

It is important to say that "fraction" here is not much in agreement with our old fraction of the set of rational numbers: division indicated between two integers, with the second of them, the divisor, being nonzero. The number is in fraction form when we write the dividend at the top, as numerator, a bar, and at the bottom, the divisor, called the denominator. If we cannot find an equivalent fraction with numerator and denominator simultaneously integers, then we have a number that is not rational. It can be an irrational number, within the set of real numbers, or it can be a complex number, with imaginary unit *i* (*i* is the only value that raised to two results in -1, $i^2 = -1$). It is in the form of a fraction, but it is not a fraction itself. The case of the golden ratio is in this category, number that can be written as a fraction, but that is not a rational number.

And how to discover the fractional form with unitary denominator? It is enough to divide the first term of the ratio (the antecedent) by the second term (the consequent). In a common fraction, such as 3/5, when we divide 3 by 5 we get 0.6. If we want to write this decimal numeral in a fractional form, just place it on 1, making 0.6/1. Every number is automatically about 1, but we just need to spell it out when we have to show emphasis or get visualization.

Let us see how one discovers the value of the golden number. A segment is taken, but to facilitate the accounts, size 1 is considered (it can be 1 meter, 1 inch, or any other unit of measurement). A little more than half, we mark a point that should make the golden section. So we have a big and a small segment in that division that we did. The whole segment is worth 1, but we do not yet know the measure of the two pieces. Then we call **x** the largest piece. How should we call the other? It is simple. Since the original segment has measure 1 and we said that the larger part is worth **x**, the smaller part is the total, 1, discounted from that **x**, which is written as 1-x.

The golden ratio point

The golden ratio is that in which the segment is sectioned so that the smaller piece (1-x) divided by the larger piece (x) is equal to this larger piece (x) divided by the whole segment (1). Our proportion will be written as (1-x)/x = x/1.

When we multiply the means, **x** by **x**, we have x^2. And the product of the extremes will be 1*(1-x). This results in the equation $x^2 = 1-x$, which, after having all the terms transferred to the first member, since it is an expression of the second degree, will give $x^2+x-1 = 0$.

The solution of this equation has a negative and a positive root. Since **x** is a segment, the negative root will not be used. The positive one is -1 plus square root of 5, all this over denominator 2. The rational approximation of the number gives 0.618 (if you, reader, remember the method of resolution, here is the invitation to take the pencil and solve the equation, otherwise, just believe the number I am presenting and move on).

What we have is that the larger segment of the section is worth approximately 61.8% of the entire segment. And between the two segments of the section, the ratio between the smallest and the smallest, by the very construction of the proportion, is also 61.8%.

Many consider as the golden number not the largest section, but the value of that section added to the whole segment, if it is of unit size, $1+x$ instead of x. This results in the number 1,618, approximately, and this is called the number phi, which is the Greek letter that corresponds to our "ph", which came us through the Latin.

Within Mathematics itself the golden ratio appears in topics as varied as the ratio for the growth of numbers in the Fibonacci sequence (1, 1, 2, 3, 5, 8, ..., in which the sum of the last two numbers always gives next) and the arm of the five-pointed star, the regular starry pentagon. If we measure end to end an arm of the star, the point at which another arm intercepts the one being measured is the golden ratio point.

Outside Mathematics, golden ratio does not only appear in architecture and industry, but also in various situations of nature. The spiral of the shell of the nautilus snail grows in buds that obey this proportion. Also the sunflower spirals, viewed from the front, follow this same ratio.

The human body also shows the golden ratio in many of its proportions, as Da Vinci implicitly left recorded in his drawing of the "Vitruvian man". From head to toe, the navel occupies the point of gold. On the face, the eyes are at the height of the greater segment of the golden section. In an English maternity hospital, at the end of the twentieth century, an experiment was made to see if infants already bring the perception of the harmonious relation of the golden number. Some cardboard masks were cut out with the eye hole at the height indicated by the proportion, others with the eyes in another position. When the mothers put on the golden masks and approached the babies, they smiled. When they went with the mask of displaced eyes, they cried.

The ratio 5 to 8 is that which between integers of a digit is closer to the golden ratio. From back to front, 8 divided by 5 is the one closest to the number phi. For in various sonatas of Mozart the ratio between the notes of the development of the theme and those of the introduction are close to the golden ratio. And in the piano, which in each octave has a sequence of 13 keys, we have a set of 8 white keys for 5 black keys.

I once asked my students at Granja Vianna, city of Cotia,

neighborhood close to Sao Paulo, to calculate on what day of the year the golden ratio point falls, that is, the date of the year that is the golden day. Quickly they found: August 13. If the year is leap year, the day remains the same, shifting only the instant, from 1 pm to 1:40 pm. What is special about this day in relation to the current year? By the golden ratio, all the missing part for the year is for the part that has already passed as the part that has passed is for the whole year.

In the solution **x** of the above equation, which gave half of the sum of -1 with square root of 5, adding this to 1, then our phi, which is 1+x, will be +1 plus square root of 5, all over 2. It is the same result, in format term, only by changing -1 by +1.

Atmosphere. It was not necessary to climb inside a spaceship, or send a powerful meteorological balloon, to get the measure of the height of the atmosphere. Arab mathematician Abu Ali al-Hassan Ibn al-Haitham, known in the West as Alhazen, who was born in present-day Iraq and lived between 965 and 1040, made this calculation around the year 1021 using a simple rule of three.

It is almost certain that his inspiration came from the method that Eratosthenes used in antiquity to calculate the circumference, and therefore the radius, of the Earth. Eratosthenes compared the shadow shifting from the edge of a dry pit in a town with the deviation, at the same time of day, into another dry pit, in another town some tens of miles away. Using a rule of three he was able to obtain, by the difference of the deviations of the shadows, and knowing the size of the arc, that is, of the distance between the two cities, the total measure of the ray of the planet, which in current measurements is 6,371 kilometers. The value he found was not exactly this, but it was very close.

To reach that value, Eratosthenes had to use the power of imagination. Alhazen, many centuries later, did not act very differently. He used the imagination and inspiration provided by scientific observation to create the method that allowed him to make the measurement.

He asked himself a question that every child also asks: why immediately after sunset does not the world around us darken? We all know that for almost half an hour the region where we are remains clear after the sun goes down. It is the duration of twilight. Alhazen sagaciously realized that the reason for this is that the Sun leaves our field of vision, but its rays continue to be reflected in the particles of the atmosphere above us. While knowing the value of the Earth's circumference, and the speed with which the points of that circumference deviate from the sunlight, he mounted a rule of three, in which the proportional incognito was the maximum height of those particles, that is, the height of the atmosphere. Let us look at the situation. If, for example, from the point where I see the Sun finish hiding on the horizon until the place where a friend of mine sees the beginning of darkness there is the distance of 250 km, then this distance is the hypotenuse (major side) of a right triangle whose height **h** is the height of the atmosphere. I take a smaller right triangle, with a 2.5 m hypotenuse, stretched out on the floor like a rope. I observe the variation of the angle of the Sun's rays in my triangle and I measure the height at this moment of darkness beginning. Then I only do the proportion with the large triangle, 250 km base. What Alhazen estimated is that the altitude of the atmosphere is approximately 100 km, which is attested today by our astronauts, including one who is my neighborhood neighbor.

By the way, the description of this procedure of Alhazen was the subject that I chose as final work for Mathematical History, when I finished my first degree.

Many scholars consider Alhazen the first scientist in history. In my estimation, the first, having invented the scientific method, was Thales of Miletus. The English author Brian Clegg, who wrote the book *Roger Bacon - The First Scientist*, considers, as the title of his work says, that this pioneer is his compatriot, the English Franciscan friar Roger Bacon. Alhazen is, yes, recognized worldwide as the Father of Optics, a subject much studied by the friar. Roger Bacon, by the way, dominated the Arabic language and repeated that, in his time, the thirteenth century, the educated man had to have fluency in three languages: Arabic, Greek and Latin.

It is reported that Alhazen was sentenced to house arrest between the years of 1011 and 1021, and at that stage he developed his work in Optics. Observing the trajectory of the Sun's rays on the wall of the house, he made several advances in the study of diffraction and refraction. The most important achievement in the area, however, is that he explained, for the first time, the functioning of vision. He realized that we can see why light rays bring the figures that are ahead of us to the lens of the eye - Isaac Newton, in the eighteenth century, discovered that what the crystalline receives are particles, which in the twentieth century were baptized as photons and had the misfortune of dividing the form of transport of light with the waveform: Einstein and other physicists concluded that light is transmitted at the same time by particles and waves, closing the polemic that came from Newton as the particle-wave dilemma.

Knowing Math, and knowing how to use it to solve problems, can be a matter of life or death. Clegg presents one more aspect of the Alhazen figure: he was recognized in his time as the world's largest problem solver. The caliph of Egypt, Al-Hakim (986-1021), learned that Alhazen told his friends in Baghdad that he had a solution to the problem of the annual floods of the Nile River. From the earliest times of the Egyptian civilization, the annual floods not only destroyed animals and crops but also displaced land from one owner to another, causing economic loss and animosity. The solution that Alhazen imagined was the construction of great dams that would serve to tame the force of the waters.

The caliph, who had a reputation of being very demanding and very cruel, demanded his presence in Cairo. He came willingly and was given him the task of studying the river and the redeemer project.

Alhazen set out on the field, with his slate clipboard and his writing stone sliver. He left up and made every possible study. The conclusion, after many weeks of work, was that the technical resources available in the eleventh century would not allow the work to be carried out for the purpose he had promised the chief of State. The fact is not uncommon, reader. In most cases of great inventions, the visionary mentally constructs the project and, when he gets to work, he discovers that there is some lack of resources, mental or material, without which the work is not complete. It happened, for example, with the computer, whose pioneering project by Charles Babbage in England began a century earlier than that of the one who, finally, presented commercially the machine to the world at the Pennsylvania State University, in 1946 (the binary machine of Konrad Zuse, in Berlin, 1934-1938, represented an important step, but not the definitive one).

Alhazen decided to return to Cairo and make his explanation to the caliph. He would say that he tried every means to work out a feasible project (factible, as many of my friends say), but he failed the task because the world did not yet provide the necessary resources, and he was not able to provide them. The verb "to fail", when it came to mind, brought him a reality that until then he had avoided glimpsing. The caliph was a very cruel man, and he did not admit that an errand that he gave to anyone had the failure as an answer. It would be death, surely.

At this point I used to interrupt the story when I told my students in classroom (at Granja Vianna, Cotia, I learned later that I was known as the teacher who taught Mathematics telling stories), asking them to think at home to see if, by chance, they discovered what was the exit found by Alhazen not to die. Rather, he reasoned with perspicacity that if he ran away he would be hunted wherever he went and would only make crueler his death. What did he do?

Arithmophobia: How to Heal the Horror of Mathematics

The students thought and in the next class they kept thinking. Never have any of them found the solution of Alhazen, which is an après-la-lettre egg of Columbus. Currently, the smart guys would go Google and find the answer, but until a few years ago, not even Google was talking about it.

Without anyone bringing the answer, I would tell the rest of the story. Obviously, there must be a true part and a fanciful part, but we cannot separate one from the other.

Slowly returning to Cairo, Alhazen set the brain to operate under a pressure he had never experienced before. Errors in strategy, and he would be beheaded.

He found the solution just before he reached the palace. Not the solution to the floods of the Nile, but to save his life. She tossed her hair, tore at her clothes, rubbed himself on the floor, and drooled on the way, using all the skills of a teacher to pretend he was crazy.

Near the palace the guards saw him and some recognized him.

- Is not this the mathematician whom the venerable caliph has ordered to study the solution to the floods of the Nile?, asked one.

- He sounds a bit, but must be some crazy man, said another.

- I still think this one is him. What do you think, Ibrahim? asked the first guard to a third.

- I agree with you. I think this is him.

- Poor thing - the second guard said, already agreeing with the other two -, he thought so much of the problem that he lost his mind.

- Let's take him to the caliph, and he'll decide what to do, first guard said.

In front of the caliph, Alhazen did not answer the questions that the monarch asked him. He just mumbled and uttered meaningless phrases.

The caliph, distrustful as every monarch must be, did not want to take any chances. Perhaps the mathematician was not mad at all, but only living through a crisis. Does not every great mathematician in studying a very difficult problem pass through this kind of setback? It was a question that the caliph could not solve.

By the Yes, by the No, he decreed the house arrest, as commented above. When the mathematician was healed, he would tell the solution to the Nile. If he did not have the solution, he would die. Ten years went by, and the caliph died. Under the new chief of State, Alhazen was released, became an Egyptian subject and lived another 19 years in Cairo, until he died naturally.

Algorithm. In the Arab world, Alhazen was aware that he owed much of his fabulous work to the intellectual trajectory of a Persian mathematician who marked the ninth century as the foremost researcher of his time. After Geometry, in the composition of the themes of basic Mathematics from the center of the Middle Ages onwards, three words began to occupy intensely the mind of the learned people: algebra, algorithm and (in Spanish) algarismo.

All three come from the name or work of that mathematician born in Uzbekistan in an area that belonged to Persia (present-day Iran). He was Muhammad Ibn Musa al-Khwarizmi, who was hired by the caliph al-Mamun, son of Harun al-Rashid, and worked at the House of Wisdom, the great academy that al-Rashid, the husband of Scheherazade, founded in Baghdad. Some scholars claim that al-Khwarizmi was born in Baghdad itself, where he spent his working life. Others say that he came from the city of Khwaresm, Khiva region, Uzbekistan, and that his name comes from the name of the city. He lived from 780 to 850, but these dates are approximations.

Why knowing al-Khwarizmi is a strong reason for liking Mathematics? First of all, because he provided a great facility for numerical studies, with the adoption, on an institutional and advanced level, of Indian numerals, which, having been irradiated from the Arab world with an intellectual center in Baghdad at that time, gained the name of Hindu-Arabic numerals. Before, the custom was to use the letters of the alphabet to represent numbers, without positional system, which made the use of Arithmetic a difficult task. The Greeks wrote an apostrophe before the letter set to indicate that they were being used with numerical role. The Romans, as we know, used the letters I (1), V (5), X (10), L (50), C (100), D (500) and M (1000), far from any symbol that could represent the cardinality of the void.

Al-Khwarizmi systematized the use of zero in positional notation, and such was the importance of his role in this history that these numerical symbols came to be called "algarismes" (digits), in French, or "algarismos", in Iberian Languages, a more simplified way of pronouncing "al-Khwarizmi". And "algorithm", where does it come from? This word is also another way of saying "al-Khwarizmi", in a more sophisticated mode than "algarismo", but meaning something more substantial than a simple numeral. Algorithm is any standardized set of steps, or procedures, that leads us to a numerical result. For example, the summing account frame, with one parcel over the other, vertically aligning the units, tens, hundreds, etc., is one of the first algorithms that the child learns in school. Another example is the resolutive formula of the second degree equation, which Baskhara invented based on a method created by al-Khwarizmi (we will see

Arithmophobia: How to Heal the Horror of Mathematics

it shortly).

The other word, "algebra", did not depart from the mathematician's name, but from the title of his work, *Al-Kitab al-Mukhtasar fi Hisab al-Jabr wal-Muqabala* (The Compendium Book on Calculation by Completion and Balancing). The word al-jabr (algebra) of the title means "restoration," or "completion". Westernized, the word "algebra" began to be used both to work with values before and after the symbol of equality, and for the art of "restoring" the bones of the human body, a type of physiotherapy practiced in large cities up at least the end of the twentieth century. At Francisco Morato Avenue, in Sao Paulo, until approximately 2000, there was a sign in a house with the expression "algebrist". It was not a service of a mathematician, but of a Japanese masseur.

Some pages above we did algebra, in the molds taught by al-Khwarizmi. This art of completion and balancing consists of passing parcels and factors to the other member of equality, respectively as subtractions and divisors. Also the powers are passed to the other side like roots, always with the use of the inverse operation. If in the first side of the equation we have an expression summed with 5 and in the second side we also have an expression plus 5, so we can eliminate this value 5 by balancing the two members: The value 5 of one side cancels with the value 5 of the other.

What method was this that al-Khwarizmi invented to solve the quadratic equation one century before Baskhara? It is an algorithm not summarized in a formula. The process name is "resolution by the completion of the perfect square method".

Al-Khwarizmi's Algebra was made extensively, because the letters representing unknowns were introduced a few centuries after him, in Europe, in the Renaissance phase, after the discovery of America. Let us use the current letters. If we have a complete equation of the second degree, as $x^2+bx+c = 0$ (if the value **a**, before x^2, is different from 1, we divide the two sides of equality by the value of **a**, so that we have coefficient 1), we try to get a perfect square within this expression. Al-Khwarizmi did this using rectangles and segments, so the root found, representing a size of segment, had to be positive. Negative root, in the drawing, was discarded.

Let us take a numerical example. Let us take the equation $x^2+6x-40=0$. How to obtain the perfect square? It is simple. We will use the formula of the first special binomial product: the square of the binomial results in the square of the first (term) plus two times the first by the second, plus the square of the second. To find out what is this "two times the first by the second" simply we divide the multiplier of x by 2. In our example, we divide 6 by 2, to see that the equation can be written as $x^2+2*3x+... = 40+...$ (The independent term, 40, was passed to the second member because it would hardly be the number to complete the perfect square, and the ellipsis indicates that we are going to find this missing value and, as we complete the perfect square in the first member, we must add it in the second member by the balancing rule.)

If the first value was **x**, and two times the first by the second gave 2*3x, obviously the second value is 3. Those ellipses have to be filled with the square of this number. We will have: $x^2+2*3x+3^2=40+3^2$. Factoring the first member, we have to put in parentheses the first term, **x**, and the second term, 3, with the whole binomial squared. In the second member, we will add 40 with 3^2, that is, 40 with 9.

Our assembly will give $(x + 3)^2 = 49$.

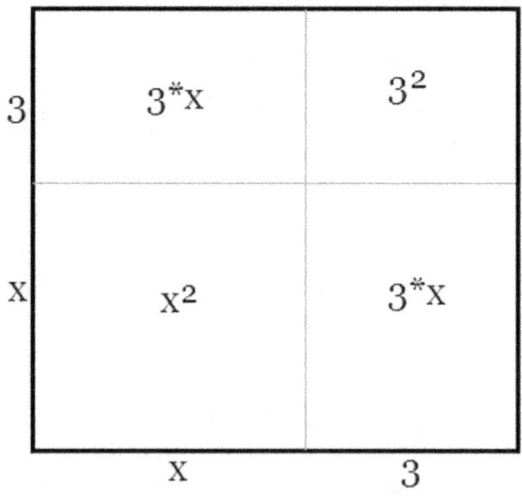

Al-Khwarizmi's perfect square

The drawing for this equation shows a square of side **x**, which we still do not know how much it is worth, plus a segment of value 3. We mark in the base x+3, joining the two pieces, and in the vertical we do the same, writing x+3 from bottom to top on the left side of the figure. We draw a segment parallel to the base at the point separating **x** and 3, and draw a vertical segment at the point separating **x** and 3 at the base. We will have a smaller square, side **x**, and, as magnification, a larger square, side x+3. In the upper part we have a rectangle with base **x** and height 3. On the right we have a rectangle with base 3 and height **x**. In the upper right corner, a small square, side 3, is formed. This square is saying that area 49 is composed of x^2 plus two rectangles of area 3x, plus a square of area 3^2. This is what we had before doing the factorization of the equation.

If the area of this large square is 49, as we saw in the second member of the equation, what will be this segment of value **x**? It is easily solved because it is known that the square side of area 49 has to be 7. If a piece is already worth 3, just do the "completion of the perfect square".

Using the Algebra that al-Khwarizmi taught us, we extracted the square root of the two members of the equation $(x+3)^2=49$. We get $x+3=+7$ (already dispensing the value -7, which would lead to a negative segment). Turning 3 to the second member, we have x=7-3. Our segment has size $x = 4$.

If we want to find the two roots, we will use the value -7, and we will have x=-7-3, which gives us x=-10. This does not enter the drawing, since there is no negative distance, or segment of negative size. Here again, the result cannot be negative if it represents area or time.

How to solve by al-Khwarizmi $x^2-10x+21=0$? First we divide 10 by 2 for finding the "2 times". Then we do $x^2-2*5x+..= -21+...$ Now we have $x^2-2*5x+5^2= -21+5^2$, which gives $(x-5)^2=-21+25$, or $(x-5)^2=4$. While extracting the square root, we have x-5=+2, or, also, x-5=-2. One result is x=+2+5, i. e., x=7, whereas the other is x=-2+5, i. e., x=3. It is easy, do you agree?

Al-Khwarizmi was also a great astronomer and a fine geographer. One of his legacies was an updated *mappa mundi*, with several corrections in the map inherited from the ancient Greeks, elaborated by Ptolemy. In this work, the mathematician of Baghdad had an auxiliary team of 70 geographers.

Crafts. If you, reader, has not yet been convinced of the opportunity to enjoy Mathematics for the value of this subject, let us talk about the professions that depend on it. First let us discuss the origin of the word. Mathematics comes from "mathema", Greek word that means learning. Pythagoras created the word, in the seventh century BC with the sense of "learning science". The other word he created for use in his school was "Philosophy", which means, as the reader knows, "love of wisdom". The subject Mathematics, in the school that Pythagoras created in Crotone, south of Italy - in the world, it was the first school for post-paideia, i. e., for post-garden, adolescents and young people -, was compounded of four areas: Arithmetic, Geometry, Music, and Astronomy.

Boethius, in Rome, little more than a millennium later, created the curriculum that was practiced throughout most of the Middle Ages from that division of areas by Pythagoras. They were the *trivium* and the *quadrivium*. The quadrivium was exactly the set of the four subjects of the Pythagorean Mathematics. The trivium was composed of Grammar, Rhetoric, and Logic.

Due to differences with Emperor Justinian, Boethius was condemned, imprisoned, and finally beheaded. At that time, in the year 529, Justinian ordered the closing of the Academy of Athens and the Museum of Alexandria. His argument was that there was no longer any need to invest in scientific research, once all the knowledge that matters is already "in the Bible". All the empire's investment in studies turned to jurisprudence in the production of the Corpus Iuris Civilis, from which there was no danger of some bewildering mathematical discovery.

Boethius had tried unsuccessfully to introduce the teaching of the Hindus numerals in Rome. The idea did not go ahead because those numerals still presented without the positional zero, so that there was not much advantage in the exchange, and his compatriots continued with their Roman numerals.

In France, in the tenth century, Father Gerbert began the European teaching of Hindu-Arabic numerals. The idea was little by little being spread and did not recede more, even because the priest became pope later, with the name Sylvester II.

But it was only in 1189, 660 years after the closure of the Academy, that Europe again had an institution with an equivalent purpose, the University of Bologna. At that time, from the Late Middle Ages, the university, restored, offered the mathematical training in the course called Bachelor of Arts. This was the graduation of Roger Bacon, for example, at the University of Oxford in the thirteenth century.

With the Renaissance, the taboo installed by Justinian was finally defeated. Great European scientists returned to their research with all their might, even at the risk of stopping at the Inquisition's bonfire, continuing what the Arabs left behind after the Baghdad collapse in 1258, invaded by the Mongols of Genghis Khan.

Today we have many solid careers based on mathematical knowledge. This is the case of Mathematics itself, Computer Science, Statistics, Meteorology, Astronomy, Physics, Chemistry, Engineering, Administration, Information Technology, Economics, Accounting, Molecular Biology and Architecture.

Arithmophobia: How to Heal the Horror of Mathematics

The careers of Computing and Statistics are within the area of Mathematics. Computing comes from the verb "to compute", which is the same as "to calculate", and from that word the verb "to count" came. "Counting" is just a corruption of "computing". Already the word "calculate" comes from "calculus", small stone, in Latin, because the ancients used abacuses of pebbles to make calculations. In the old days Statistic was seen only as a use of Mathematics for purposes of government and administration, but the works of Carl Friedrich Gauss, in the nineteenth century, and Nikolai Kolmogorov, in the twentieth century, have shown that it is one of the chapters of rigorous Mathematics.

As stated above, Pure Mathematics began in 1848 with the launch of George Boole's book *Mathematical Analysis of Logic*, which later he expanded as *The Laws of Thought*, 1854. Bertrand Russell locates at this stage the separation between Pure Mathematics and the applied sciences that until then were considered as an integral part of Mathematics, such as Physics, Astronomy and Navigation.

What has happened is that since then, studies that involve time and matter, such as Physics, have come out of the scope of Mathematics, which, as my friend Odilon Luciano says, deals with stopped things, or dead things. (Odilon Luciano, for those who do not know him, is the earliest teacher in the history of the University of Sao Paulo, USP, having started his undergraduate studies as a colleague of mine at the age of 14, along with 17. Later, the USP have banned from enrolling student under the age of 16, so that his primacy cannot be dismantled.)

When in computing we speak of processing time, the word "time" in the context refers to the number of steps, a purely mathematical idea. This is not clock time, although it is not forbidden to measure the processing time of a program with hours, minutes and seconds.

Does restricting the study of static things diminish Mathematics? On the contrary. It contributes to making it simpler and therefore richer and safer. At the beginning of the nineteenth century the French positivists classified the sciences according to the following criterion: From the simplest and general to the more complex and particular. This has resulted in the order: Mathematics, Astronomy, Physics, Chemistry, Biology, Anthropology. Anthropology would be a comprehensive human science, which would include sociology, politics and others. Over time, this human science was called Social Sciences, having Anthropology as one of their branches. Psychology, which was not yet a laboratory science, was rejected by these positivists, and it was not until the end of the century, in 1879, that Wilhelm Wundt, in opposition to this prejudice, installed in Leipzig the first laboratory of the subject, which, strictly speaking, is a biological science. Many of the views of these philosophers are now seen as outdated, but the view that Mathematics is the simplest and most general science has always been confirmed, never doubted, except by people who have very little affinity with the subject. One cannot confuse "simple" with "banal", however, because it demands dedication and attention.

It is important to mention that Boole's book had the explicit purpose of transforming Aristotle's Logic, until then a philosophical discipline, into a chapter of Mathematics. In the book he shows, step by step, that all processes used in the analysis of propositions can be incorporated into what he called Propositional Calculus, or "Algebra of Logic". Boole was a friend of Augustus De Morgan, who had been translating some results of Logic into Algebra, but Boole was the one who showed that there was no stone on stone left in the "discursive logic" building. The whole Logic of the philosopher of Stagira was transformed into Algebra. That would have given a great scare in the Macedonian, that was not very affectionate to Mathematics. He could scarcely imagine that within Mathematics we would have not only the whole Logic, but also chapters such as Disaster Theory, Chaos Theory, and Theory of Fractals.

Progression. In Number Theory, a progression, or sequence, is a set of numbers that we take, at random or under some criterion, represented by values separated by a comma, or semicolon. It can be a set of infinite numbers, or a finite set, as long as it is ordered, that is, it is formed by elements that do not change their position ad libitum. Thus, the sequence (4, 2, 7) is very different from the sequence (4, 7, 2). These are two finite sequences. Examples of infinite sequences are (0, 1, 2, 3, 4, ...), progression of natural numbers, and (8, 1, 153, -10, 47, ...), a random sequence. If we want to represent an unordered numeric set, we write the values between braces, not parentheses.

Among the sequences obtained from a criterion, or a law of formation, the most used and behaved are Arithmetic Progression and Geometric Progression. An Arithmetic Progression (AP) is the sequence in which each term, after the first, is equal to the previous term plus a fixed number, called "step", indicated by the letter **d** ("common difference"). A Geometric Progression (GP) has almost the same law of formation, only by observing that each term is equal to the previous one multiplied by the fixed number, which is now legitimately called "ratio", or "common ratio", indicated by the letter **r**. An AP can be increasing, constant or decreasing, whereas a GP can be increasing, constant, decreasing or oscillating. An example of AP is

(3, 5, 7, 9, ...). The step here is the number 2, as we just have to discount 5-3 to have this value that is added to each number to get to the next one. An example of oscillating GP is (1, -3, 9, -27, 81, ...). What is the ratio there? It is enough to divide -3 by 1, to obtain r = -3. This is the value that multiplied each term produces the following.

Sequences are of undoubted utility, especially in working on computer programming. But the subject came here as a subterfuge to get back to the greatest researcher on Number Theory in the whole history, the German Carl Friedrich Gauss (1777-1855). When Napoleon Bonaparte, also a professor of Mathematics at the Polytechnic School of Paris, took his troops to the east with the unfortunate purpose of conquering Russia, in the phase of crossing Prussia he asked the soldiers to spare the city of Brunswick, explaining that "the greatest mathematician of all time resides there".

Behind this episode it is a poignant story about the situation of women in the history of science, even after 14 centuries of the martyrdom of Hypatia of Alexandria. Sophie Germain, a great French researcher, used to exchange letters with Gauss in high mathematical discussions. She would have entered the Polytechnic if that college accepted women. That was not yet the case. There was a young man known to her, Antoine-Auguste Le Blanc, who was studying there but did not follow the course well. He gave up, then, and moved from Paris, something the teachers did not know. She then took the assignments for him, solved them and put them in the teachers' bins. They corrected them and, in response, praised the young man's progress. So she grew up in science, posing as a man. In correspondence with Gauss, she used the same resource, of course.

At the army's departure, she asked one of the officers she knew well, Pernety, to intercede with Napoleon in favor of Gauss.

Pernety did not resist and wanted to meet Gauss himself. He told him of Sophie Germain's commitment to preserving the German's life, for fear the soldiers would do with him what Marcellus' men in ancient Rome did to Archimedes. There, other Sicilians saw when a soldier attacked him on the back with a spear while he was on the beach, absorbed, studying a figure casually formed in the sand, and did not respond to a question asked by the soldier, even though the general had warned him before that he wanted Archimedes alive. Gauss only found strange the name of his protector, Sophie Germain. He said that he did not remember anyone he knew in France who called Sophie. Pernety then spoke of her mathematical work, and Gauss realized. She was the young Le Blanc, with whom he exchanged so many academic correspondences without knowing that he spoke to a woman.

Sophie Germain died in 1831, well before Gauss, but was born only one year before him, in 1776. They were almost the same age and, thus, shared the same scientific concerns.

As a child, student Carl gave his teacher a hard job because he was very quick on assignments and always had time off, while his classmates suffered to accomplish the obligations in class. Carl was the example of the true hyperactive student, who does everything ahead of schedule and starts talking and nudging others. Thomas Alva Edison, later in the United States, was another example. He spent only two months attending school, because his mother, knowing that he took many raps in his back for hyperactivity, took him out of the classroom to continue home his education. Obviously he had already learned the letters and was already literate in those two months. (Do not do this at home, reader, because we are no longer in the nineteenth century.)

Some authors say that Gauss was eight years old at the time of the episode that we will see here. Others claim that his age was six. We do not know who is closest to the truth, but the facts advocate for eight, as we will see.

The problem was in Mathematics class, because in the rest there was consonance between the Gaussian rhythm and that of the others. In a dictation, for example, all students follow the master at the same time. The teacher decided to find a way to keep Gauss busy while the others did their chores, each in his own vagaries. He imagined he had found the way out.

When everyone was doing the proposed exercise and Gauss obviously had already finished, the teacher introduced the novelty.

- Carl, I have a specific task for you.
- You can order, sir. I'm ready.
- You will add the sequence of integers from 1 to 1001.
- As well? Add 1002 numbers?
- It would be hard for another student, not for you.

Gauss went to his desk and the teacher breathed a sigh of relief. Now the boy had something to do for the rest of the school day.

The teacher's quiet, however, lasted very little, a few minutes if a lot.

Gauss reasoned in the way that comes next. If we have to add only the first three numbers, 1, 2 and 3, the sum of the three will give the same value as the triple of their arithmetic mean, that is, $1+2+3$ is the same as $2+2+2$ - the fact that Gauss already knew the notion of arithmetic mean weighs in favor of those who bet on the age of eight.

Gauss did another test. If the sum goes up to 5, does the idea work?

We will have 1+2+3+4+5. The average, which coincides there with the middle term, is 3, because it is 15:5. But in a sequence like this, there is a greater advantage: the arithmetic mean is the mean of the two ends. In the latter case, it is the average of the two numbers 1 and 5. As the average gives 3, simply multiply this value by the number of terms, which is 5, to reach the total sum. We have 3*5=15. In the previous case, it was the average of 1 and 3, which gave 2, and then multiplying the value by 3 one comes to 6. Then he saw that it was enough to take the two ends and divide by 2, multiplying this average by the quantity of numbers. If we want to add from 1 to 9, we do the arithmetic mean of the two edges, dividing 1+9 by 2, and the result 5 multiplies by the total number, which is 10.

Ready! Gauss "summed" those 1002 numbers in a few minutes and brought the result to the teacher. Did he sum? Not! He used the discovery he just made: The sum formula of the first AP terms. He simply added 1 with 1001, got 1002, divided the value by 2, to get the average 501, and by this value he multiplied **n**, the quantity of numbers, that was 1002. That gives 502.002.

Finally, if the first number of an AP is **a** and the last one we want is **k**, the sum Sn of these first terms, that is, from **a** to **k**, is given by [(a+k)/2]*n. If the first term is 5, if the step is 3 and we want to add up to 38, we must first find that the total **n** of terms is n=12 (if necessary, we use the general term formula of an AP, k = a+(n-1)*d) and then apply the formula. The sum will be [(5+38)/2]*12. When the sequence is of integers, always the value of the parentheses, or the value of outside, one of the two, will be divisible by the denominator 2. We get [43/2]*12, or 43*6. The result is 258.

After Gauss's death, people discovered he was writing a diary. It was not a diary of ordinary amenities and events, such as what he had lunched or with whom he went to the theater, but a diary of scientific activities. On almost every recorded day there was a mathematical fact, almost always a new discovery.

Attractiveness. There are those who dedicate themselves to Mathematics by discovering something divine in it, as it was the case of Pascal, and there are those who see in it the most pleasant professional activity among those we have available. This was certainly the case of Simeon Denis Poisson (1781-1840), to whom we owe important works in the areas of Electricity and Probability. He said, "Life is good because of two things: doing math and teaching math."

We can say that people are bound to life through divinity or through

Mathematics, which is, according to Galileo, the language with which the deity communicates with us. The atheists and agnostics have, therefore, in Mathematics their greater reason to continue living. The theists and the deists who cultivate Mathematics have a double reason to live.

Is it very difficult to keep up with this science? No, if the student and the teacher understand that there are those who embrace it as a craft of life and there are those who need to learn from it the basic knowledge for everyday life. It is very comfortable to click an icon on the computer screen, or press this icon with the finger on the screen of the tablet and, observing the action in progress, know that this occurs through a mathematical function, knowing well the concept of function. Either the user knows how the mathematical operation of the machine occurs, or all the technology in front of him comes as the magic of some great manipulator. It is not by chance that the number of victims of persecutory delusional disorder, collectivized, the old paranoia of the conspiracy, is increasing lastly.

"Whatever your difficulty in Mathematics, I assure you mine is greater." When Albert Einstein said this he was not confessing to be a weak student in the subject. He was warning, implicitly, that the problems that you have to face using Mathematics are much simpler than those that he encountered in his studies of General Relativity Theory. Yes, major problems require great tools. For those who are not scientist, engineer, economist, or something like this, the necessary Mathematics is that of basic education.

And what about an atheist who does not want any courtship with math? This has Logic as a good field of study to devote, not to run away from Mathematics, but to use it as a topic that nobody can conscientiously dispense with - the one who has quarreled with Logic is the one who is sick to his head. The problem is that Logic is easy and familiar to those who have undergone good training in Geometry and Algebra. Otherwise, it is a swamp of quicksand: When the student imagines that he is knowing everything, he is doing everything wrong. In order to formalize Logic, Aristotle did not have time to realize how much of that was indebted to the notions of Geometry he learned from Plato.

9. The strength of Geometry

We have many remedies against the disease of arithmophobia, but the most powerful in terms of vaccine, once it can reach the entire school population, is Geometry. With the guarantee of a good learning of Plane Geometry on the part of the young, the arithmophobia evaporates like the measles evaporated among the children.

In the book "Projects for Brazil", Jose Bonifacio presents on a page the 20 reasons why Portugal did not do well in science. One of these 20 reasons, left to the end of the list, is the finding that Portugal "gave no importance to Geometric Drawing".

Perhaps aware of this, Benjamin Constant Botelho de Magalhaes, who developed the curriculum for the Pedro II School in Rio de Janeiro, a curriculum that gradually was extended to the entire Brazil, with an emphasis on the basic sciences, something rare in the world (other countries have a more science-oriented curriculum, but not for the entire population).

Elimination. Everything had been walking slowly, without a hitch, until, in the Municipality of Sao Paulo, the lawyer Janio Quadros, in his second passage for the chair of mayor, from 1986 to 1989, abolished the subject Geometric Drawing of the curriculum of the junior high school, which was called Fundamental Teaching II. Two or three years later, another lawyer, Governor Fleury Filho, did the same with the curricular matrix of the state schools of the State of Sao Paulo. The coincidence of the law degree in both cases is not indicative that the law firm throws to one side and Mathematics to another. Lawyers who respect and cherish Mathematics, such as tax lawyers, labor law lawyers and those from any other legal area, do much better in the cases they defend than those who are victims of arithmophobia. If you are a lawyer and you know that your opponent has a horror of math, you have discovered the weakest of all his weaknesses! And how did those two, taken by arithmophobia, become chiefs of State? Now, because the majority electoral system in Latin America is still tragicomic.

In both measures, in the Municipality and State, it was argued that Geometric Drawing is a chapter of Mathematics and should be treated by the Mathematics teacher, not having a specific matter to study the subject.

As the fashions of Sao Paulo, capital and inland, are quickly followed by the other regions of Brazil, the standard became the elimination of the subject Geometric Drawing.

Five or six years after this horrendous loss, I began to receive students with severe symptoms of the illness in High School. At the time when the matter was cultivated in all elementary schools, one or another student presented resistance to Mathematics, almost always because of lack of mastery of one or another topic, mainly the addition of fractions, but no case was pathological. If the student had the mental conditions to learn languages, natural sciences or Geography, in a short time he made peace with Mathematics. From 1996 onwards, the reality has changed.

Does have Geometry not been taught by the mathematics teacher, according to the governmental plan? Not to say that this is totally untrue, a range of 1 in 50 teachers works a bit of Geometry in his math classes. It is as if the Sabin vaccine, against poliomyelitis, was applied only in 1 in 50 day care centers. The gain would be almost negligible. It would be to lead Doctor Sabin to die again, just as the measure of Fleury Filho and Janio Quadros once again killed Master Euclid of Alexandria.

The experience of eliminating content arguing that it would be embedded in another was nothing new. This was how the military regime, in the High School reform of 1971, abolished the Music of Villa-Lobos, taught in the subject Orpheonic Song. They wrote that it would be incorporated into the new subject Art Education. Music at school died from there, just as Geometry practically died in the early 1990s. But would not it be the case to simply guide teachers to do that work? No, it is not so easy. No need justifies harming academic freedom. In order for students to learn Music, a specific subject matter must exist. In the time of the quadrivium the teacher of Astronomy was not required to teach Music. The same thing happened with Geometry. Each had his obligation.

Treatment. If the family wants to vaccinate its children against arithmophobia and they are in a school system that follows this pattern of disregarding Geometry, it can solve the problem by hiring private tutor of the subject. If the student does not have to make test (in the short term) about the subject, it is difficult to convince the student to dedicate himself to this, but it is not luxury or snobbery, otherwise a preventive medicine. Later, without having studied the subject in school, the student will have to face it, either in the entrance exam of the Technical High School, or in the college entrance exam.

The strategy of hiring a tutor for Geometry is not something unheard of in history, certainly. The best known case is that of Philip II of Macedon. He hired Aristotle as preceptor of his son Alexander, who would later become Alexander the Great. Aristotle taught Philosophy, Rhetoric, Logic, Physics (at that time Physics was more about Biology than Mechanics), etc. But Aristotle had no affinity with Arithmetic and Geometry. He fulfilled his obligations as a pupil of Plato, even though the famous phrase "do not enter if not geometer" was written on the portico of the Academy, but he was never charmed by the beauty of Mathematics, as can be seen from his work. Thus, Philip decidd to add another preceptor, expert in Mathematics. He contracted geometer Menaechmus, another student of Plato.

One of the advantages of learning Mathematics through Geometry, at least in case of heirs of Iberian culture, is that the learning disguises itself as something outside Mathematics. Geometry is the core of basic Mathematics, as General Architas of Tarentum told Plato to convince the Athenian to found the Academy. But if it is taught as "Geometry", "Geometric Drawing", "Euclid", or whatever name you want to give not remembering the word Mathematics, the student learns more sophisticated Mathematics, as is the demonstration of theorems, without realizing this. When he least expects, he is positively and emotionally committed to Mathematics. Is not it deception? Now, if it is proven that placebo cures tummy ache, will we reject the treatment because it is not allopathic? But it is not the case of placebo use that of Geometry. There are experiments showing that certain children who do not accept to eat broccoli, eat the product with satisfaction if the mothers ask before they close their eyes. If the color of the broccoli scares, the substance matters. We abstract its color. If the name Mathematics frightens, use the substance, which begins with Geometry.

Geometry, unlike Arithmetic, functions as the visual alphabet of Mathematics. When we begin learning to read, we must master our 26 Latin letters. The Greeks, 24 letters. Knowing by heart the 26 letters and having mastery of their functioning in the formation of words, we are literate. Knowing the meaning of each word is another step that never ends, although some madmen out there have been saying in recent years that literate is who, besides reading, knows the meaning of any written word. This type of people has not been well literate, nor does he know the dictionary, where we always find words that we never saw or did not remember existed.

Since the late nineteenth century, no one has been able to make any new discoveries in Euclidean Geometry. It is assumed that the terrain is exhausted, and there is no more to be gained in it. Far from understanding that this is a reason for abandoning the area, we must be aware that it is the alphabet, the code from which Mathematics becomes meaningful for us.

This code is complete, unlike Arithmetic.

If one dreams of being an architect, one must first appreciate the learning of Geometry. If he wants to be a plastic artist, idem. If he wants to be an industrial designer, value Geometry even more. If he wants to know how to argue well in court, as a jurist, or in the rostrum as a politician, he must be aware that learning Geometry Theorem Demonstration methods is the best way to present his exemplary performance as a professional.

But if the young person intends to pursue a medical career, he may think that Geometry, like all Mathematics, will only serve as a way to win the race for college. He is wrong. If in his primary and junior high school course he has had the opportunity to handle, to the minimum necessary training, instruments such as ruler, square, compass and scissors, all used in Geometric Drawing, he will be a surgeon who will hardly miss the cut in the patient's body. Geometric Drawing, as well as Music, gives the student the opportunity to train both hands. A doctor who has studied Geometric Drawing in elementary school will be in a much higher condition than he who only used his hand to write texts in his elementary school years and, not being a southpaw, only had the opportunity to train his right hand, even if he has passed by the artistic drawing.

Approach. At what age should the child start studying Geometry? Well, when he starts literacy. In kindergarten, before the school itself, which usually begins at the age of six, the child spends much of his time practicing artistic drawing, while drawing lines and painting. Arriving at regular school he already has some domain of the use of the hand to make tracings. Along with literacy, he should start learning Geometry by drawing and distinguishing segments, rectangles, triangles, and circles.

When the child begins his addition learning, he can add segment measures, rather than just making abstract accounts. To sum 15+28+17+8, he can remove these values from a polygonal line, a sequence of four segments whose measures are these. The question is: how long is the polygonal line?

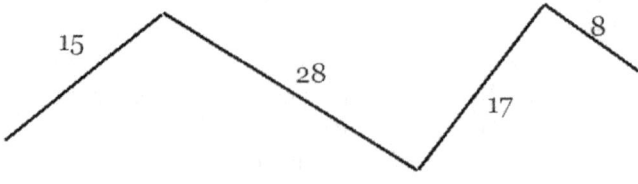

Several polygonal lines can be displayed for the child to calculate the size. And he can also make sums to find the perimeter of a rectangle, or an irregular quadrilateral.

Once he gets used to calculating perimeters, the figures can be used in subtraction. For example, we present a triangle with sides 18, 21 and (?), With the information that the perimeter is 54. To approximate more the practical world the idea is that these measures are all in centimeters. If the sum of the three sides is worth 54, how to find the measure of the side that has interrogation? The child must perceive, or must be helped to realize, that by summing the two given sides he will have a subtrahend, which, operated with minuend 54, will give the rest, which is the answer to the interrogation. It will make 54-39, getting 15.

When he reaches multiplication, a given problem may be a polygonal line with segments of the same size. If we show a line consisting of 8 segments, each measuring 13 centimeters, the child will multiply 13*8, and get the product 104.

He can also get the perimeter of regular polygons. If it is a triangle, he multiplies the side by 3; if it is square, by 4; if it is a pentagon, by 5; and so on.

And once we are in the multiplication phase, we introduce the concept of area. The child will calculate the area of a rectangle by multiplying the dimensions base and height. We can also find the area of the triangle by multiplying the base by half the height. We can write the formula extensively: base*height/2. Later on the child will learn that this becomes $A = b*h/2$.

And how to use the division account in the geometric figures? Doing, as in the case of subtraction, the inverse operation, from the totals. We give a regular hexagon with perimeter 42. What should be the value of each side? The child will count and will see that he has 6 sides. He simply divides.

In the fourth or fifth grade, or even in the third, when operations with decimal numerals come, the segments present measurements with dots. We can ask for the perimeter of a quadrilateral whose sides, indicated in the figure, measure 7.3; 12.6; 9.8 and 11. The child will do the operation 7.3+12.6+9.8+11.0, putting dot under dot, and will get the answer 39.7.

Looking at what he is calculating is a powerful way to get the child to become familiar with the numbers, losing his fear.

If Mathematics is the science of learning, according to the intention of Pythagoras, when he invented the word, Geometry (meaning "measure of the Earth") is the instrument of clarification of this learning, the right front leg of this tamable quadruped pet, which is the Pythagorean quadrivium.

Games. I have written and published a few years ago a small volume on numerical games, in the form of a fiction story. There, I recommend that the numerical games themselves should not take much of the classroom time, so that the feeling of returning to kindergarten, which should prevail for the well-being of the child through the fun, is not installed. In regular school, the time of the hobby is the break time. But of course the teacher can use some class space to show how students can practice numerical games by proposing to do so outside the normal classroom.

An easy and enriching type of game is Numeric Expression. The student constructs a 20-sided dice, which is an icosahedron, and numbers these faces twice from 0 to 9. He and his play partner take even or odd fingers to decide who starts. Whoever starts, throws the dice and takes the values from the upper face, to copy them in the outline of the expression, as follows.

$$9 + [\ldots (\ldots + \ldots) + \ldots] - \ldots$$

Student X, who started the game, fills the spaces indicated by the ellipses with the values taken in the icosahedral die - the value before the first parenthesis is obviously a number to be multiplied, because parentheses means "times". He resolves the resulting expression and writes the total. Then student Y, his partner, does the same task by filling in the five numbers in the diagram, copied on another sheet of paper - the value 9 at the beginning is to ensure a positive total. Whoever gets the highest score wins the match.

School. The Ministry of Education absorbed that initiative of the rulers who decided to obscure and undermine the teaching of Mathematics by dismantling its support, which was the learning of Geometry.

There is no other way to restore education and to heal the young people affected by arithmophobia if not the rescue of Geometry in school.

While top management does not accept to correct the error, the local systems can begin recovery. The curriculum of the third triennium, the high school junior course, counts on six weekly classes of Mathematics, along with six weekly classes of Mother Language and smaller numbers for the other subjects. Now, by giving the Mathematics teacher the "obligation" to develop his Arithmetic and Algebra topics without neglecting those of Geometry, school will not leave the trapdoor in which it was cast. But regional boards of education, municipal secretariats, state secretariats, and even school units, on their own, can organize a distribution of lessons that circumvent the insidious obstacle.

As the number of weekly classes is even, six classes, the school unit (if not some system that is above) can decide, as its own project, that each class of students will have two Mathematics teachers, one of Algebra (that incorporates Arithmetic), with name Mathematics, and another one of Geometry, each with three weekly classes. A school unit with eight classrooms, making up 48 weekly math classes, does not deliver 24 classrooms from four classes to one teacher and another 24 from the other four classes to the other. It assigns to the two teachers classes in the eight classrooms, each ministering three by class. If they prefer, the teachers can divide the assignment, being only one with Geometry and the other only with Algebra. But they can also switch: in classroom 7A John teaches all three Geometry classes and Lucy teaches the three Algebra classes, reversing roles in 7B. It is more convenient and practical, however, for John to take Geometry in the eight classrooms, and Lucy to take in all eight Algebra only. The priority when choosing fits, as is customary, the teacher with the highest score of time and titles.

At the moment of delivering the bimester (or trimester) averages, once the materials are not yet separated by the system, but only by the school unit, John and Lucy calculates the grade of each student, merging the two marks in a arithmetic mean. If Andrew obtained 6 in Geometry and 9 in Algebra, the average in both subjects will be 7.5, which will be transformed into 8 if the system requires a whole mark - if it were, for example, 7.3, the rounding would be down, getting 7, according to the universal rule.

Advertisement. The Mathematics teacher, both with Algebra and Geometry, should, whenever possible, remind himself that he should do in the classroom an advertisement of his subject, convincing students of the importance and necessity of studying it. And he should always remember old topics that have fled from the memory of students and which they need at the moment. The teacher uses the contents often, but the students are in another situation.

An alert that must be made to the authorities of the school unit, as well as to the teachers and parents, is that, if the students have not yet studied Geometry, and have in the seventh grade, for the first time, the subject given by a specific teacher, the marks from the first bimester will be low, perhaps much lower than the math scores until then. The motive is the language, with a profusion of Greek vocabulary, and a new type of learning, which is that of theoretical spatial organization. In the second bimester, however, the grades will be equal to those of Algebra, and from the third bimester will be higher, if the students dedicate themselves.

The school unit that takes the decision recommended here will soon see its students delivering spectacular results in the Mathematical Olympiads.

10. The etiology

The time has come to reveal the fulcrum of the whole problem. It is now.

When I graduated from Mathematics, I organized with my colleagues the I Mathematical Education Week. As the Institute building was new, with only one wing functioning, we had no auditorium, so we borrowed that from the Oceanographic Institute. That place was where we hold the lectures and seminars of the Week. I had read a magazine from Canada, talking about holding a colloquium there for the same purpose. That is where the idea came from to reproduce the event in Brazil.

We invite the main authors of textbooks of basic education, such as my friends Osvaldo Sangiorgi, Luiz Barco and Ruy Giovanni, my master Scipione Di Pierro Netto, Professor Ubiratan D'Ambrosio, from the Unicamp, and other important names in the area.

We appointed the colleague Mario Takazaki to make the table with the guests, and I made a point of staying in the audience, because at the table the organizers lose mobility. On the last day I asked for the word, from the place where I was. In my speech I dealt with arithmophobia. Professor Ubiratan said that he hung for the term Mathophobia, as a version of Math Anxiety. I explored the vision I had worshiped until then, which was to be that of a Latin problem, of the Latin-speaking countries. Some in the audience grumbled, probably disagreeing from my point of view. The auditorium was crowded, and Professor Samuel Pfrom Netto, from the Institute of Psychology, who had been sitting in the background, without my having seen him, stood up in my defense. He said:

- The teacher is right, because in my wanderings around the world I do polls, asking in the classroom if the students like Mathematics. In Denmark, one student from each classroom raises his hand to say he does not like it. In Chile, only one student raises his hand, but to say that he likes it.

The audience complied with Professor Samuel's argument. As for me, I kept investigating, trying to discover the origin of the disorder.

Refining. Years later, teaching in a junior high school class in Butantan, city of Sao Paulo, I received the students' complaint that they were taking low grades because the exercises I put in the test were very high. It was the beginning of the school year and we were still in the stage of remembering the fundamental subjects, which they did not dominate. I was accustomed to my classes at Granja Vianna, which I had been following for years. In this group of Butantan, each student came from a different school and, if one knew some Algebra, ten knew almost nothing.

At home, I thought, "They may be right, and my workouts may be demanding a lot."

In the next class I warned that I would do a test without my exercises, but only of questions taken from books. I did this and, in fact, the class average improved somewhat, not substantially. I am telling the episode here because I made several discoveries there. One was that my exercises were not difficult, but the textbooks are very easy (maybe for a business issue).

The other discovery gave me a scare.

For they to know that I had not included any of my exercises, I wrote in the proof, at the end of each question, the name of the author, book and page from which the exercise had come. What was the surprise? At the time of setting up the test I did not see anything special, but when I corrected it, I saw that the names of the authors, of the five questions, were all Italian, beginning with Sangiorgi.

I got up and went to my bookshelf. It was not just those five authors. Almost all of the Mathematics textbooks I had were from "oriundi" authors. I do not remember at this time any that did not have Italian origin, although it existed.

I was immediately reminded of Fibonacci, Luca Pacioli, Archimedes of Syracuse, Galileo, and Cavalieri. I turned my mind to France and recalled Viète, Descartes, Pascal, Legendre, Lagrange, Cauchy, Condorcet, Galois, Poincaré, Dieudonée, and many others.

So I searched mathematicians of Iberian culture in my mind. I remembered my professors, most notably Newton da Costa, the greatest logician in South America, author of the Paraconsistent Logic, which did not have this name yet when I had a class with him. The Poles gave the denomination later.

Although Lusophone and Hispanic in America have contributions to Mathematics, only recently has there been a recognition, with the Fields Medal (maximum prize of the area) being presented to Professor Artur Avila, IMPA-RJ, in 2014.

The problem of arithmophobia, I realized that day, is not Latin, but only of the Iberian culture. It remained to continue the investigation to locate the beginning of everything.

Muslims. Over time, I began to incorporate extra-numerical tasks into the evaluation items, to compose the students' grades at the end of the bimester period. For example, I sometimes demanded they wrote on some subject. This went in as a plus or as a fraction of the mark, ten percent, for example. In a given year, 1996 or 1997, I asked, at the end of the first semester, that the students write a text speculating on the origin of the rejection to Mathematics by the Lusophones and Hispanics.

I came to the semester recess with a bunch of almost 400 texts to correct. On my return to school I warned that nobody had discovered the source of the problem, but that their effort was very worthwhile. They did not find out, but they pushed my conclusions to a definitive solution. They talked a lot about history, about Brazil difficulties of formation as a country, about the connections with Portugal.

What I realized in those days is that the question is really related to History. Not from the historical facts themselves, but from the interference of these facts in the way of life of people, bringing consequences that persist for centuries.

The problem was in colonization, but not in our ownly. The reading of the texts of those adolescents reminded me that before we were a colony of the Iberian, they were colony of the Arabs. Here was the key.

In the year 711 it began the occupation of the Iberian Peninsula by the Moors, the Muslims who already dominated the Arabian Peninsula and the entire Maghreb, which is the north of Africa.

Only half a century later, in 761, al-Mansur the Victorious, al-Mahdi's father and grandfather of Harun al-Rashid, founded Baghdad, which in Persian means "divine gift".

Between one enterprise and another, the Arabs attempted, using the Iberian Peninsula as their spearhead, to take France, when they were defeated at the Battle of Poitiers, in 732. Without this victory of Charles Martel, on behalf of France, the whole of Europe almost certainly would have been taken by the Muslims. That is why the Roman Catholic Church gives that warrior chief the title of "Hero of Christianity".

Without the intended domination of France, the Arabs chose to use the financial gains flowing from the Maghreb and Iberian Peninsula to strengthen their initial base, which resulted in the founding of Baghdad around the ruins of Babylon.

The seat of the world caliphate settled there.

With the founding of the House of Wisdom by Harun al-Rashid, the city began its glory in scientific production, with great mathematicians, astronomers, historians and geographers, as well as in fiction literature, with the monumental legacy of the tales of "The Thousand and One Nights".

All was well with them until a great warrior, trained to be the right arm of the chiefs of armies in Spain, received an atavistic "call". Rodrigo Díaz de

Vivar (1048-1099), taken by his father to train in the military arts with the Muslim leaders, presented a growth that was far above normal in his warlike activities. He received from the chiefs the nickname El Sid, the Lord, that is, "Lord of War", a name later spelled as El Cid, or Cid the Campeador.

The role of Jimena, his fiancee, represented in the cinema by Sofia Loren (1961, with Charlton Heston doing El Cid and directed by Anthony Mann), in his position by the War of Reconquest, is not very recognized until today, but, surely, a great change is always based on decisive influence. At one point, El Cid turned the game over. With Christian ancestors, far back in the past, what sense did it make that he was now lending his intelligence and his strength to maintain Baghdad's domination over the West? He joined his men, those who agreed to stay with him, to defend the territory, and took the region of Valencia, in 1094.

That was the initial push for the resumption of the Iberian Peninsula by the Christians in battles that would continue for another century and one half. Several independent kingdoms were being created, such as Aragon, Castile, Navarro, Leon and Portugal. Many were joined later to form Spain, but, as we know, Portugal stood apart.

One of the great gains of this phase was the formation of schools of translators, who poured from the Arabic into Latin many stories that the Muslims had translated and maintained from the Greek. Among these entities were the School of Toledo and the School of Amalfi. The most important documents were just mathematical books.

After great and numerous battles, the Arabs were finally expelled from the Iberian Peninsula, in the year of 1249, in conflict in Portugal.

Nine years later, Baghdad fell under Hulagu, grandson of Genghis Khan. It was founded 50 years after the Arab invasion to the Iberian Peninsula and resisted only nine years after that half-millennium colonization period.

Doctrine. All war is based on deception, Sun Tzu wrote in his book *The Art of War*, of the fifth century BC. It can be imagined that it might be different, but war is already the result of a human weakness, mounted in ignorance, then the fact of being based on lie, or misunderstanding, adds no more tragedy to what it already represents itself.

As the Reconquest War was basically a religious war - and almost every war is - the deception was related to doctrine. It is not a distortion of Catholic teachings, nor of Muslim rites and beliefs, but a doctrinal addition on the part of those who needed to harness forces to change the situation, and this was the Christian side.

Arithmophobia: How to Heal the Horror of Mathematics

For almost 300 years the advantages that the Arabs showed about the culture and religion they brought convinced the Iberian. At one point, the rebellion took place and that indoctrination of three centuries had to be undone. The truth was revealed, but that was insufficient. Then within the population people spread some innocent lies, which will not bring serious consequences up front. That is what they thought.

In Brazil we have no rejection of the kibbeh, the tabbouleh and the Arab surnames. In the Iberian Peninsula a politician of Arab surname does not have the facility to rise in the race as he has in Latin America. Those more than two final centuries of the colonization counted on an intense campaign of rejection of the Arab culture. The words stayed, the minarets remained, but some victims were innocent items. The most wronged of them was the digit.

The name is an "algarismo", but the object is Hindu. For the Iberian warriors, it mattered the name, "algarismo", which came from al-Khwarizmi, Arab thing.

This is very painful, but it is the fact. It was not the Indians, but the Arabs who brought those numerals. Before, the Iberian lived together with the Roman numerals. Taking them back was part of the reconquest, as well as returning to the lands. To resume something, it is necessary to demonize the substitute.

Today it is important to tell young people, first, that we have very little to do with that story. Even the Portuguese have little to do, because all those Arabs who dominated the Iberian Peninsula are dead, for many centuries. Second, that the numerals we use are Hindus, having only been brought by Arabs. And third, that these numerals are powerful and endow with great power the people who dominate them and are not afraid of them.

Napoleon Bonaparte said that we can recognize the potential of a country for its investment in the study of Mathematics. Hitler, knowing this, dictated for his scribe Rudolf Hess to write in one of his two books that the way to make the "Aryans", who he thought were the Germans, to dominate other peoples was to forbid others to learn advanced Mathematics. A Lusophone or a Hispanic who continues to have math horror is fueling the feeling that Hitler wanted to lead him to cultivate. And you, reader, should not imagine that someone with clearer skin would have a bigger chance to dodge the racism of the false German. Among the machinations that he dictated to Hess, knowing that the accusation that they "killed Jesus Christ" was insufficient for his purposes, was the justification why the Jews should be hated: They were guilty of bringing the

Negroes into the Rhine Valley, with the aim of "weakening the German blood".

In Brazil, for an almost continental journey of Professor Sangiorgi, a consensus was created among Brazilian Mathematics teachers, that the use of a calculator in class should only be allowed when the adolescents were mastering the basic accounts. That is, as long as the kids think they will need an electronic calculator, they should not be allowed. With the invasion of cell phones, Sangiorgi's fight is going downhill. President Macron has sanctioned in France in 2018 the prohibition of cell phones in the classroom. Here, Governor Jose Serra had, years before, sanctioned a device (State Law 12,730/2007) determining the prohibition in the State of Sao Paulo. Now Brazil needs a national ban, as was the case in France.

In Portugal, teachers require the use of a calculator in class. We do not know yet how it will be now, with the opposite example of France, because the prohibition of the cell phone is not only by dispersion in social networks and videos, but also by the use of the calculator as a crutch.

Counting on the fingers is sound, because it does not swell memory and reasoning. Instead, it helps the brain to develop. With the calculator the student has no chance to reason, because the result comes ready, once we provide data. If any data is entered in error, the result will be wrong. And the student will pay for his ignorance.

By the way, how does the Japanese fingerboard, cited above, work? Let us see. The student needs to master the multiplication table well up to number 4. Our table of fingers works from the number 5.

We use our fingers as fives. If in one hand we want to express the number 7, we count, from the little finger, 1, 2, 3, 4, 5, without lowering any finger. We continue the second corner from the little finger. When we count 6, we lower that finger. When we count 7, we lower the contiguous ring. What we have now is a hand with three raised fingers and two lower ones, representing the number 7 (so we only represent from 5, because we would not be able to distinguish the values 1, 2, 3 and 4. See that the number 5 is represented with all the fingers raised, and the number 10, with all the fingers lowered.

Now the cat's leap appears. Raised fingers will be counted as units. Fingers down, like tens. If we want to do 7 "times" 6, we put in the right hand the number 7, as described above, with two fingers lowered, and in the left hand we counted until the 6, lowering only the minimum finger. We will have two fingers down on the right hand and one on the left hand. They are three fingers down, forming three tens, the number 30. They are three raised fingers in the right hand and four fingers in the left hand. We multiplied these fingers: 3*4. The result 12 we add with the tens, and we will be with 30+12 = 42.

Now you do, reader, to practice, the 9*8 count. The total of lowered fingers should be seven, forming seven tens. The rest is easy. It is adding down tens and multiplying raised units. Continue, reader, practicing other accounts of the table from 5 to 10, even if you already have it all by heart.

11. Redemption cases

Magnitudes. By counting on the fingers, or using only the brain and the pencil, the student will learn to evaluate order of magnitude of numbers. With the constant use of calculator, million, billion, quadrillion, sextillion, all this has the same value, something great, which the student never sees.

By the way, the logic error that fair calculators present is perceived by very few people. It was made by engineers, to meet a specific demand, which does not include junior or high school students. When we want to get a percentage, already incorporated as an increase, we have to calculate this percentage and then add it to the base value. For example, what will be the new price of a $ 800 value product that gains a 5% increase. We calculate 5% from 800, arriving at 40. Then we add up to the base value, 800. The result is 840. For sophomore students of High School, or more, we teach them to multiply 800 by 1.05 (this last value means 100% plus 5%). In the fair calculator, the user types 800, presses the "+" symbol, and then type 5 and press the symbol "%". Using Logic, he is adding 800 with 5 hundredths, which gives 800.05. The result of the calculator, illogical, gives 840. It assumes that when the user hit "%" after typing 5, he wants to add that result to the value 800. It is a dangerous joke that in the circles of education scholars we call "mathemagics". Let us flee from it!

Of course, when the student has fun with his cell phone he never thinks about the creators of this electronic piece. It does not occur to him that every time he presses an icon with his finger to perform some action, this has resulted because of mathematical functions that the programmers used in the assembly of the system. No sound, no image, no result, of any kind, will come up without a mathematical work behind.

- What do I have to do with it?, asks the young man.
- You have much to do, his teacher says.
- I don't want to be a programmer, neither for cell phone nor for computer.
- Nor do I want you to become one of them, but if you start discarding the possibility of the future, you may run out of the future.
- I doubt it. What I want as a career does not depend on Mathematics. I want to be a soccer player.

The teacher goes on to explain to the student that Brazil has won five men's soccer world championships when the country still cultivated a minimum of Geometry in school. Without any knowledge of Geometry, the Mathematics alphabet, players make goal without having any idea why. And without knowledge the results are mediocre.

- Is that why Brazil stopped winning the Cup?
- That's the main reason.
- And how do we win back?
- There are two possibilities: Brazil returns to giving importance to Geometry, or the competing countries destroy their own education, as Brazil did. Then we will all be on an equal footing.
- This is very sad.
- Of course, but start by doing your part, studying Geometry, Arithmetic, and Algebra.

Reviews. A lady gave testimony to a newspaper some time ago explaining why she developed the problem of "math anxiety".

By the ninth grade she had always performed well in the subject, but when she entered High School, her teacher was a weak and very demanding instructor, and she took a class with him during the three years of the course. It was with this teacher that she found herself suffering from arithmophobia.

Now, she will never know that she was deceived at the fundamental level. It is safe to say that what she had there as Mathematics was not Mathematics, but only a shadow of the matter. If the Mathematics teacher had fulfilled the official determination to teach the chapters of Geometry in a well-crafted way, she would be saved from developing the disease later. The assignment left in the hands of the junior high school teacher is misplaced, because what works is the matter Geometry given separately, but when they did the merger the professorship did not complain.

In High School it would not matter the didactic of the teacher, because what the student has to learn then is only a complement of Mathematics of the ninth grade. Everything is just a list of formulas and a set of concepts and techniques that, for those who have learned Geometry and Algebra of the fundamental course, are very easy. Take the example of multiplication of matrices. The great difficulty of the students in this subject, and that causes them to take low marks, is the domain of use of signs in the multiplication of integers. If the student came with this training from previous years, everything is just a fun little game. Chapters like trigonometry and analytic geometry require prior knowledge of Geometry, certainly. The High School teacher may try to fill gaps in those who have been duped at the fundamental level, but he will have little time to do so while teaching the regular program chapters.

Aware of this problem, I and a fellow teacher decided to start our first year classes in High School with a week of review of the fundamental course with emphasis on geometric aspects. In the first year it was so, but in the following school years we saw that it was too short time, and we decided to extend the period to one month. Later, for one and one half month. In these weeks, we "reviewed" all of the Algebra of the fundamental course using the theorems of Geometry. I have written the word "revised" in quotation marks because for the vast majority it was all new.

To review proportions, we used the Thales Theorem and the similarity of triangles. In order to study the first degree equation, we used several other theorems, such as the inscribed angle, the circumscribed quadrilateral, the sum of the internal angles of the triangle, the definition of complementary angles, and so on. For percentages, we used lengths and areas. For systems of equations, we used the theorem of the external angle and that of the angles of two secants. For the equation of second degree, we used point power and the Pythagorean Theorem. In order to have complete quadratic equations, the measurements did not come only with the unknown **x**, as in the books of the ninth grade, but with expressions of the type x+1, 2x-3, etc., which, after being squared, result in complete polymomies of second degree.

A girl who got very low marks in the first semester, something like 1 or 2, started to have only high marks in the second semester, from 8 to 10. I sat down with her to have a conversation.

- You are a good Mathematics student, so why were your grades so low in the first semester?

- I could not get good grades at first, sir.

- Why not?

- Because I knew almost nothing about Mathematics. So I decided to pay attention, write down everything, do my best, and have confidence in me.

- The effort paid off.

- No doubt. The whole revision of Geometry and Algebra that you passed opened a light on my head.

Some time later I became school coordinator. Disagreements with the bureaucracy of the Department of Education caused me to temporarily leave the unit. I spent three years at a nearby school. When I got back, the classmates had reorganized the program of the subject, adjusting it to what the bureaucrats of the Teaching Regional Directorate had always tried to impose. The review phase had been abolished. The school unit, which was the best among state schools in the State of Sao Paulo, attested to in several indicators, became not only equal, but inferior, to the units of the region.

Arithmophobia: How to Heal the Horror of Mathematics

Polyhedra. It was already the height of increasing of the arithmophobia when an emblematic case occurred in my classes. It was a second-year High School class and there was a student in it who did not even look at the board, but who really liked Philosophy. He always had some old thinker's book in his hand. He did not attend his first year in the school unit, and therefore did not go through our review of Geometry.

One day I was about to start a new chapter, before I wrote the subject on the blackboard I warned what I was going to do.

- Today we will start the theme Polyhedra of Plato, the Platonic Solids (later I wrote the title on the blackboard).
- What do you mean, master? Plato's Polyhedra?
- Exactly.
- Is he the Greek philosopher? From Athens?
- For sure. He was a great geometer.

The student, who sat at the bottom of the class, at that moment moved to a desk ahead of the class.

I won a student who was lost. Did I Win? No. Plato won.

He began to focus on matter. Getting grade was no longer problem for him.

Spark. Geometry has this appeal, of being the subject of Plato's study, and of many remarkable brains. But sometimes the birdlime to tie the student to Mathematics is in topics that we do not value highly, diluted that are in the middle of so many subjects that we have to teach.

I once received an eighth grade student transferred from a very expensive private school. Already in the first days I realized the reason for the transfer. His father was an engineer, a resident of a mansion on the same street as the school, but the boy, unlike his father, did not give the due importance to Mathematics. He was a student who did not like to study. He did not do homework. For a whole bimester he messed up the class.

At the beginning of the next bimester, entering a new chapter, I warned: Today we are going to start the chapter on Systems of Equations. Then I wrote this on the blackboard.

He stopped making noise, stopped disturbing, and came to ask what the subject meant. He liked the word "systems". Son of an engineer, it was common term in the conversations that he listened at home. I suppose that was the reason for his interest.

Since then, he was doing all the tasks I was presenting. One day he came to talk to me saying that his father asked me if I could indicate more exercises. I attended. Months later, in a meeting of parents and teachers, his father praised my work in front of the History teacher, who was the coordinating teacher of the boy's classroom, and asked the teacher to communicate this to me.

In another class, also in eighth grade, that same school year, a girl who did not like to do lessons or follow the explanations, was enchanted by the method of al-Khwarizmi. When I finished the factorization exercises, I said that I would show how to solve the equation of second degree by completing the perfect square. I made an example on the blackboard. She did not shut off even a second of that explanation. I wrote some exercises and she did them.

I decided to stay longer on the subject because it is algebraically rich and makes the student get the basis for new adventures in Algebra.

The girl made all the equations I passed, over fifty, always hitting. Obviously, I showed how to check if the answer was correct (if do not you remember, in $ax^2+bx+c=0$, with the two results in hands, if they exit, the product equals **c** and the sum equals -**b**). I thought: This is a student who was won for math. Naive deceit.

I went into a new chapter and waited for her to respond positively. She was again the same girl as before. I am talking about a time when the elementary school student was given the power to learn or not, according to his taste. If the score was zero at the end of the school year, the promotion for the next grade was guaranteed, without any mishaps and without the student carrying a pending. What was not guaranteed was a reasonable way of life for those who threw away their most precious time for learning, which is that of childhood and adolescence.

Unthinkable. Another case was that of a girl who arrived in a classroom of mine, transferred from another school, there for the second month of the school year. It was a seventh grade class.

She sat in the front row, near the teacher's desk, and always did the chores. She was an exemplary student, with blue marks every bimester.

At the last meeting of parents and teachers of that school year, her mother made a point of going to the room where I was, which was not the girl's, to talk to me.

That mother said her daughter had always done poorly in math. In every year the problem was repeated: Low grades, rejection of matter, lack of courage to do tasks, and all those difficulties that children present before the subject when they have not yet been won.

I had a hard time to believe, but I did not have to doubt that mother, who was truly grateful to the teacher who finally made her daughter lose her fear of Mathematics, according that report. I told her that with me she always performed well and I could not guess she had the problem of rejection before.

I think today of the reasons that made the girl reject Mathematics until she was my student. I see three possible hypotheses. One, there was for some reason an involvement of the pathology, from an early age, and it lacked someone who encouraged her to escape from it. Another, by chance, she had in front of her, to that moment, teachers who treat very badly Mathematics itself (certain time I left a school unit to come to teach closer to home, I was replaced by a teacher who required students to memorize everything, with contempt for reasoning and deduction, causing as a result that people who only took 1 and 2 with me would take 9 and 10, while good students began to have a mediocre mark). The third hypothesis, which we always try to do not to be true, but which, unfortunately, may have been it, is the racist attitude. Approximately 1% of teachers cultivate morbid racism, whether explicit or not. The girl was a beautiful mulatto, more to black than to brown. (Please, reader, despise the prejudice against the word "mulatto", because that case happened in Brazil. I assure you that this recommendation to avoid the word "mulatto" was a racist thing. We must maintain the understanding that we Brazilian are black, white, mulatto, American Indian, brown-skinned, Mamelukes, brown, Semites, Slavs, Indians and Orientals, striving to ensure that racist contamination does not poison our coexistence, and hence prevent us from permanently eliminating the pockets of racial indoctrination that swarm here and there.)

For Mathematics, there is no preferential skin color. This exists for the racist individual, not for science. Blacks, mulattos and Indians can win the Fields Medal just as they can win the Formula 1 World Championship. If today the Medal seems distant, it is because the scientific world began in northern Greece, migrating to Arabia and then to Western Europe and North America, having not yet taken root in Africa. Japan and China are already integrated. Great Indian mathematicians have already shown their potentialities for many centuries. The populations of sub-Saharan Africa will also be integrated. Hitler knew that in all nations there are able people, and that is why he had the plan to dominate by boycotting and evading information. Yes, learning comes from empathy. In front of a racist teacher, the child who feels rejected is blocked. Racism is one of the greatest barriers against human progress.

Suppose that a contagious, fast-spreading, lethal disease emerges, and that the only person with the entire potentiality to decipher it and beat it is someone from an ethnic group rejected by racists. Suppose a racist teacher denied that person access to a minimum degree that allowed him to reach the channels within which he would show his ability. Let us suppose that nobody knows about this racist teacher, and that not even the injured person was aware that he was vetoed by him, thinking that his low grade was due to his incompetence. Now, a mere melanophobic citizen, feeling demigod, and trying to act as such, led to the decimation of humanity by pure racism. Let us think, reader, how much the countries of the Americas have been squandering in human potential by leaving in the middle of the path capable people whose phenotype displeases racist people who have the power to decide others' ways.

I have always been a supporter and enthusiast of the Mathematical Olympiads. I very much supported the work of Professor Shigeo Watanabe, whose son was my friend in college. I only talked to the teacher once, but I should have had more contact with him. He was the creator of the Sao Paulo State Mathematics Olympiad. Today we have, next to the Brazilian Olympiad of Mathematics, the Brazilian Mathematics Olympiad of the Public Schools, created by Mrs. Iole de Freitas Druck, a friend of mine of many years. This Olympiad has been contested by about 20 million students each year, and has revealed potentialities in the most unexpected regions. If a teacher is in the right to block the path of a studious and skilful student, in other ways this student can be taken to the place to which he is entitled.

Summary. Towards the end of the twentieth century, I elaborated a list of the 20 necessary techniques that the students should have brought from the previous course so that, if they identified the topics absent in their learning, took care to fill them. The 20 techniques are the following ones.
1) Operations with integer numbers (zero in quotient, rest of division, etc.)
2) Divisibility criteria (by 2, 3, 4, 5, 6, 7, 8, 9 and 10)*
3) Prime numbers (Sieve of Eratosthenes and decomposition)
4) Calculation of LCM by decomposition in prime factors
5) Calculation of HCF by successive divisions
6) Operations with fractions
7) Operations with decimal numerals (dot under dot, adjustment of houses, etc.)
8) Power of fractions and integers
9) Priority of operations and parentheses
10) Proportion, rule of three and percentage
11) Set of signs in parentheses, fraction bars and modules.

12) Distributive property and operations with monomials
13) Algebraic factorization (taking factor in evidence, perfect square, etc.)
14) Simplification of expressions and algebraic fractions
15) Replacement in formulas (numerical value) and systems of equations
16) Canceling in expressions and systems
17) Simplification of radicals and rationalization
18) Resolution of equations and inequalities of first degree
19) Resolution of equations and inequalities of second degree
20) Resolution of irrational, biquadratic and literal equations
(*) Divisibility by 7: Remove the last digit and double it; subtract it from the truncated number; do this until the result has only one digit; if its absolute value is 0 or 7, the number is divisible by 7. Example: For the number 1792, we do 179-2*(2)=175; 17-2*(5)=7; 1792 is divisible by 7.

If the necessary language rigor is neglected at the moment the student writes mathematical expressions, however much his numerical correctness is working, he compromises his arguments if he wishes to question the teacher's correction of his tests. Often, by not paying attention to writing, the student says exactly the opposite of what he intended. Here are the 12 most common types of mistakes that students make, sometimes even by trying to mimic practices poorly commended by others. Training and care should be taken to avoid the 12 faults below, which great crowds of students commit.

1. *Putting* a unique denominator under the equation and, worse, canceling it - write (2x+1)/3=(2-x)/3.
2. *Cutting* with the denominator a *plot* on the fraction bar – in (4 + x)/4, do not cut 4.
3. *Canceling* denominator in any number, out of equation-inequality: 7/2 is not 7.
4. *Disrespecting* the order of operations – always avoid adding before multiplying.
5. *Multiplying* number by an indicated sum without putting the parentheses.
6. *Dividing* number by expression that has value 0.
7. *Confusing* the function with its argument – sen30° is not 30°.
8. *Inverting* the sense of implication turning it into equivalence.
9. *Writing* the implication symbol (→) between numbers (it is only

between sentences).
10. *Relaxing* on the proportionality of segments, e. g., in graphic axes.
11. *Using* the symbol "=" between different values – e.. g., sum is not average.
12. *Considering* that there are parentheses where they do not exist: -3^2 is not $+9$.

The student who completes the ninth grade must also have mastery of the seven properties of the real number powers. Their names are (1) product of powers of the same base, (2) division of powers of the same base, (3) power of power, (4) distributive of the exponent in multiplication, (5) distributive of the exponent in division, (6) power of negative exponent and (7) power of fractional exponent. Taking values **a, b** in R- (0, 1); **x, y** in R and **n, p** in Z-{0} (Z: Integers), these properties are as follows:

1. $a^x * a^y = a^{x+y}$ (ex.: $3^m * 3^4 = 3^7 \Leftrightarrow 3^{m+4} = 3^7 \Leftrightarrow m+4=7 \Leftrightarrow m=7-4 \Leftrightarrow m=3$)

2. $a^x / a^y = a^{x-y}$ (ex.: $2^m / 2^3 = 2^5 \Leftrightarrow 2^{m-3} = 2^5 \Leftrightarrow m-3=5 \Leftrightarrow m=5+3 \Leftrightarrow m=8$)

3. $(a^x)^y = a^{x*y}$ (ex.: $(5^2)^3 = 5^{2*3} = 5^6$)

4. $(a*b)^x = a^x * b^x$ (ex.: $(3*5)^2 = 3^2 * 5^2$)

5. $(a/b)^x = a^x / b^x$ (ex.: $(2/7)^3 = 2^3 / 7^3$)

6. $a^{-x} = 1/a^x$ (ex.: $5^{-2} = 1/5^2 = 1/25$)

7. $a^{n/p} = \sqrt[p]{a^n}$ (ex.: $\sqrt[5]{3^m} = 3 \Leftrightarrow 3^{m/5} = 3^1 \Leftrightarrow m/5=1 \Leftrightarrow m=5$)

In a dozen chapters I have tried to explain the problem of mathematical horror, to show how it has become more pronounced in recent times, to bring the necessary psychological and didactic remedies, and to decipher the origin of the disorder, since the understanding of a defect will only be complete when we discover what gives rise to it. I have tried to use as few notations of Symbolic Algebra as possible, once the idea here is not to teach Mathematics but to discuss what makes it a bogeyman for many people.

I also sought to scrutinize the teaching and learning of Arithmetic in the final years of the primary course and the passage of this to the next level, the junior high school, for it is at this stage that the child tends to break his friendly relationship with the subject, what occurs when in the curriculum there is no highlight for Geometry.

The serious damage that societies have by cultivating this disadvantage is not new and is well known. There are many studies in the way of the solution and the one I present here is obviously one more, which is not to say that it is any, because here the roots of evil are unearthed, what allows us finally to implement an efficient prophylactic policy.

As far as the moment of learning Mathematics is concerned, the family, the school and the student should avoid anticipating much and also the opposite, to delay too much: Nine-years-old children are not yet ready to learn demonstration of Geometry theorems, while learning addition of fractions should not be left for adult learners, age 30 or older. An adult can learn this, but what the child learns in play the adult learns in the suffering, with difficulties that cause pain. Obviously, studying the topics at the right time is the most advisable. We have to think of the four degrees of detachment of the spirit, from Benedetto Croce. According to him, the development sequence of the human intelligence is this: Aesthetics, Logic, Economics, and Ethics. In the basic school, Aesthetics is organized through Geometry and Music. As for Logic, it comes through Algebra, which comes after Arithmetic.

Let us take the vaccines, and let us not get discouraged!

@cacildo

www.ingramcontent.com/pod-product-compliance
Lightning Source LLC
Chambersburg PA
CBHW071403220526
45469CB00004B/1150